Wood Pellet Heating Systems

Wood Pellet Heating Systems

The Earthscan Expert Handbook for Planning, Design and Installation

Dilwyn Jenkins

SERIES EDITOR:
FRANK JACKSON

from Routledge

First published 2010 by Earthscan

2 Park Square, Milton Park, Abingdon, Oxon, OX14 4RN
605 Third Avenue, New York, NY 10017
Routledge is an imprint of the Taylor & Francis Group, an informa business

First issued in paperback 2020

While the author and the publishers believe that the information and guidance given in this work are correct, all parties must rely upon their own skill and judgement when making use of them – it is not meant to be a replacement for manufacturer's instructions and legal technical codes. Neither the author nor the publisher assumes any liability for any loss or damage caused by any error or omission in the work. Any and all such liability is disclaimed.

This book was written using principally metric units. However, for ease of reference by readers more familiar with imperial units, the publisher has inserted these in the text in brackets after their metric equivalents. Please note that some conversions may have been rounded up or down for the purposes of clarity.

Notices
Practitioners and researchers must always rely on their own experience and knowledge in evaluating and using any information, methods, compounds, or experiments described herein. In using such information or methods they should be mindful of their own safety and the safety of others, including parties for whom they have a professional responsibility.

Product or corporate names may be trademarks or registered trademarks, and are used only for identification and explanation without intent to infringe.

ISBN 13: 978-1-84407-845-5 (hbk)
ISBN 13: 978-0-367-78754-7 (pbk)

Typeset by Domex e-Data, India
Cover design by Yvonne Booth

A catalogue record for this book is available from the British Library

Library of Congress Cataloging-in-Publication Data
Jenkins, Dilwyn.
  Wood pellet heating systems : the Earthscan expert handbook on planning, design, and installation / Dilwyn Jenkins.
     p. cm.
  Includes bibliographical references and index.
  ISBN 978-1-84407-845-5 (hardback)
  1. Stoves, Wood–Handbooks, manuals, etc. 2. Wood pellets–Handbooks, manuals, etc.
  3. Dwellings–Heating and ventilation–Handbooks, manuals, etc. I. Earthscan. II. Title.
  TH7440.P45J46 2010
  697'.04–dc22
                                                                    2009052605

Contents

List of Figures, Tables and Boxes

Figures

Tables

Boxes

Acknowledgements

Additional research, design and writing credits

Brian Horne, Teilo Jenkins (diagrams), Rachel Ceri and Tess Jenkins.

General acknowledgements

Frank Jackson, Michael Fell, Claire Lamont, Andrew Boroughs (Organic Energy Company and ÖkoFen UK), Christine Öhlinger (Energy Agency of Upper Austria, ESV: O.O. Energiesparverband), Herbert Ortner (ÖkoFen), Christian Rakos (ProPellets, Austria), Powys County Council, Bob Beaumont, Chris and Sue Cooper, Andy Bull (Severn Wye Energy Agency), Andrew Stewart (Coed Cymru), Chris Laughton (The Very Efficient Heating Company), Jacinta McDermott, Yvonne Jones (Blazers Wood Pellets) and Tom Owen (Forestry Commission in Wales).

List of Acronyms and Abbreviations

ACCA	Air Conditioning Contractors of America
BMS	building management system
BTU	British thermal unit
CE	European Conformity
CHP	combined heat and power
CIBSE	Chartered Institution of Building Services Engineers
CNRI	Canadian Natural Resources Institute
DHN	district heating network
DIN	Deutsches Institut für Normung (German Standards Institute)
DTI	UK Department of Trade and Industry
DOE	US Department of Energy
ESV	O. Ö. Energiesparverband (Upper Austrian Energy Agency)
EU	European Union
FSC	Forestry Stewardship Council
FTC	US Federal Trade Commission
GJ	gigajoule
HETAS	Heating Equipment Testing and Approval Scheme
HUD	US Department of Housing and Urban Development
IEA	International Energy Agency
J	joule
kW	kilowatt
LPG	liquid petroleum gas
MOB	material other than biomass
MWe	megawatt of electricity
MWh	megawatt-hour
NGO	non-governmental organization
Nm^3	normal cubic metre
NO_x	nitrogen oxide
ONORM	Österreichisches Normungsinstitut (Austrian Standards Institute)
PFT	PerFluorocarbon tracer (gas)
PLC	programmable logic control
PSSR	Pressure Systems Safety Regulations
SI	International System of Units
TW	terawatt
WSE	written scheme of examination

List of Acronyms and Abbreviations

1

Introduction

What is wood pellet technology?

Wood pellets are a robust new fuel for space heating, hot water provision and, at a large scale, the generation of electrical power. They are a highly versatile wood-based product designed to compete with fossil fuels on convenience, performance and price. Overall, the technology encompasses production, supply and use of both wood pellet fuel and wood pellet combustion equipment.

The burgeoning 21st century wood pellet industry is led by Sweden, the USA and Canada, the latter alone producing round 1.3 million tons of pellet fuel in 2008. Europe produced around 9 million tons of pellets in 2007, led by Sweden and Austria. An innovative source of renewable energy, if the source is sustainably managed woodland or a clean waste wood, these small pellets are recognized as a virtually carbon-neutral fuel. Wood pellet technology has evolved rapidly since the early 1980s when it was first developed in Sweden as a practical and efficient way to use the timber industry's waste wood and the country's indigenous forest resource. Since then, the issue of climate change has reared its ugly head and wood pellets now provide a cost-effective, carbon-neutral and convenient alternative to fossil fuels such as coal and fuel oil for space heating and, increasingly, renewable electricity generation in the form of combined heat and power (CHP) where the power can be sold to the grid and the heat distributed to buildings and/or used as process heat.

Wood pellets are dryer and much denser than comparable fuels such as wood chips. This makes them free-flowing, easier to store and cheaper to transport. Their compact nature means they yield a relatively high heat. Similar in form to animal feed pellets – tubular, usually between 6mm (0.24 inches) and 12mm (0.47 inches) in diameter and 10–30mm (0.39–1.18 inches) long – they are usually tan to green-brown in colour, depending on the constituent material (see Figure 1.1). They are easy to ignite and produce both less ash and lower emissions than other wood fuel applications, and the ash they do produce is easy to handle. Their size and shape also allow for optimum air flow, which is important for maximizing heat outputs. Their small size enables loading via automated boiler-feed mechanisms. Some pellet space heating systems are so well automated that they need no human operator or maintenance beyond loading the winter store, pressing a button to start the system, then removal of ash six months later. This level of convenience competes well with that offered by oil and gas boiler systems.

Pellets themselves can, in theory, be made from a great many different organic and synthetic materials. Given this, ensuring quality control over the fuel stock's source, moisture content, admixtures and contamination levels is

Figure 1.1 *Wood pellets*

essential for the end-user and the industry as a whole. Only such guarantees will maintain confidence in the pellet product, assure the longevity of the combustion equipment and, of course, stimulate further demand from contented end-users and the good reputation earned by the technology. Most wood pellets are made from clean wood waste or clean forest sources such as sawmills, joinery factories and well-managed woodland thinnings. The supply of pellets to the end-user is already established in many regions of Europe and North America and at several levels. Handy 10kg (22 pound) bags can be picked up at some corner stores and petrol stations for domestic space heaters, while at the top end, big trucks deliver to larger end-users by blowing pellets down a tube straight into a fuel store; in Stockholm, boats are used to transport pellets to a CHP station.

A guaranteed regular supply of quality-controlled pellets has been fundamental to the uptake of pellet combustion equipment. Dedicated stoves for room heating, boilers for larger spaces and even CHP pellet technology (see Chapter 4) have emerged in less than 30 years. Biomass co-firing is finding its place in large heat and power stations. The Netherlands burns pellets in a pulverized coal plant and similar developments are in their early stages in the USA. Space heaters and boilers on the market these days range from state-of-the-art, designer models to more basic and less expensive systems. Various levels of automation are available. Assuming quality control over feedstocks, three main factors give pellets an edge over other wood-based and carbon-neutral fuels: reliability in automated systems (even at a small scale); the lower cost of transportation; plus higher and more predictable heat outputs and a higher degree of temperature control.

Related technologies: log burning and wood chip

Of course, intentionally burning wood has a very long history. *Homo erectus* used it for warmth some 500,000 years ago and we *Homo sapiens* have used it for processes such as cooking and pottery making for around 10,000 years. Wood-fired underfloor heating was used by the Romans, Vikings and Anglo-Saxons. Before the fossil fuel age began in the early days of the industrial revolution, pretty well the entire human race depended on wood as its main source of heat. Water and wind power were developed to some degree, but wood was the dependable heavyweight source of energy until coal was taken up on a large scale during the 19th century. Now, as we approach the point of peak oil production and fossil fuel reserves are diminishing rapidly, the value of wood as a renewable fuel resource has been recognized once more. Not only do trees absorb carbon as they grow, but, managed properly, forests and woodlands are a theoretically endless source of wood.

Logs are the traditional way to use wood as a fuel. Today, there are scores of modern wood log stoves and boilers, many designed for pretty high efficiencies, low emissions and, compared to older stoves, much greater convenience of use. Apart from pellets, wood chips are the main alternative tree-derived fuel to logs. Medium to large boiler systems tend to use chips or pellets rather than logs, mainly because these forms of wood fuel are easier to use in fully automated systems. In domestic properties, or for single rooms, pellet technology is far superior to both wood chip and logs. Chip boilers don't work so well at this small scale. Logs take a lot of human handling, while pellets can be automated even at the smallest of scales.

The technology for burning logs has certainly improved over the millennia, but they are usually burnt in relatively simple ways: open fires, stoves, stoves with back boilers and dedicated boilers. Recent advances in boiler designs and controls have led to the manufacture of automated log boilers marketed for the larger house or small to medium commercial buildings. Generally, these boilers require stacking with logs just once a day. Workable automated log supply mechanisms would be very difficult to design and expensive to build, which is perhaps the main reason for the tendency towards using chips or pellets in larger installations.

Chip technology has advanced significantly further. Several large Scandinavian cities depend on wood chips for their heat and power, and some city centres in Finland even use chip-based heat to warm pedestrian pavements in downtown shopping streets to help avoid ice and snow underfoot. In the USA, pellets are more common at the small space heating stove level and, where they are compatible, in larger co-fired power stations. Space heating boilers, from small to large, are also being taken up at an increasing rate. The UK is a relative newcomer to wood chip heating, but it already has scores of schools and other local authority buildings enjoying the benefits of medium to large modern, automated wood chip boilers. Generally, though, wood chip combustion technology is designed for the upper end of the scale (wood chip boilers are available from about 30 kilowatts (kW), but are most reliable from about 200kW and over) – so unless you've got a really big house, large commercial building or a group of properties close together, you're better off looking at wood pellets or logs as an energy source. Wood chips flow well with automated

Table 1.1 *Wood heating systems compared*

	Log stove	Log boiler	Pellet stove	Pellet boiler	Chip boiler
One room	■		■		
Average house	■			■	■
Large house		■		■	■
Larger commercial or public building				■	■
Properties on district heating network				■	■
Requires boiler space		■		■	■
Requires significant wood fuel store space		■		■	■

Source: Dilwyn Jenkins; developed from an earlier version in the *Wood Fuel in Wales* booklet, Powys Energy Agency (2003)

boiler-feed mechanisms and most medium to large combustion units will cope with even relatively wet chip. Since wood chips rarely achieve lower than 25–40 per cent moisture content, this is a big advantage. In energy terms, pellets are presently more expensive than chips in Europe, with pellets at €10–13/gigajoule (GJ) and chips at €4–7/GJ.

However, in most automated boiler feeds, wood pellets flow even better than chips. Pellets are also much dryer and, by weight and volume, give off significantly more heat. In today's market place, pellets appear to be the most versatile of all wood-based fuels. They are appropriate to almost all requirements, regardless of ownership type (independent or cooperative) and building size. They're also the most convenient to handle and offer the greatest number of advantages when used in automated systems. For users with ready access to free or relatively inexpensive logs or wood chips, the versatility of pellets may not be a relevant factor. But for those who have to buy in fuel, wood pellets arguably provide the most effective and efficient technology. Table 1.1 compares the different types of wood heating system.

Examples of installed systems

Wood pellets are technically viable for most heating applications, so there's a wide range when it comes to the scale of systems already installed. For space heating in an average room, manufacturers – particularly in Austria and Finland – produce stylish and competitively priced room heaters. As with fossil fuel combustion the room heater and boiler systems described below – like all wood pellet installations, however small – require connection to an external flue or chimney.

Pellet stove

Make: Envirofire
Output: 4–12kW
Cost: £1000 approx.

An Envirofire pellet stove was installed during 2000 by Coed Cymru, the non-governmental organization (NGO) dealing with woodlands in Wales, as a basic room heater in a large exposed office space in windy Mid Wales. With easy access to a range of dry wood matter and its own small-scale pelletizing machine, Coed Cymru did not have to buy in all its pellets. A single-skin flue was exposed for 5m (16 feet) within the room to maximize heat use. On an annual basis, its stove consumes around 650kg (1433 pounds) of pellet fuel at a rate of about 1kg (2 pounds) of pellets per hour. Maintenance requirements have proved simple, simply regular emptying of the firebox and the occasional emptying of the ash pan, plus removal of fly-ash that accumulates near the base of the flue. At the time of writing, the users reported that this stove has operated well for nine years. Their only real issue is that its output is too small for the large office space where it is installed.

Figure 1.2 *Envirofire room heater (4–12kW)*

Source: Coed Cymru

Pellet boiler

Make: Perhofer Biomat (7–22kW)
Output: 7–22kW
Cost: €11,000

This domestic pellet boiler was installed in a newly built and highly energy-efficient Austrian home in 1997. The boiler is fully automated and linked to a solar thermalwater heating installation. Ash removal is manual. An early project, even for Austria, poor-quality pellets caused problems in the initial stages. However, with pellets that conform to the European standards – Deutsches Institut für Normung (DIN), Österreichisches Normungsinstitut (ONORM) or the more recently proposed CEN European Standard (CEN/TS 14961: www.pelletsatlas.info/cms/site.

Figure 1.3 *Perhofer Biomat boiler (7–22kW)*

Source: Perhofer

aspx?p=2550) – the boiler system works very well and consumes only around 4.5 or 5 tons of pellets a year. High energy efficiency measures incorporated in this new house have helped to minimize fuel requirements. A domestic pellet system, with a larger output (17–50kW), was installed in a bigger house in Poland during 2004. This example uses around 11 tons of fuel annually but cost significantly less at around €5000. Reflecting equipment prices on a more mature market and the increasing cost of fuel oil, this project had a return on its investment within five years.

Figure 1.4 *KÖB 540kW boiler*

Source: KÖB

Medium-scale pellet boiler

Make: KÖB
Output: 540kW
Cost: €98,000

This pellet boiler was installed in 2000 to replace a gas boiler in an Italian hotel. No significant civil works were required and the heat distribution system was already in place. Requiring around 50 hours of annual maintenance time, the system adequately heats up to 10,000m² (107,639ft²) of hotel space through a mini district heating network (DHN) of insulated pipes. Annual fuel use is around 100 tons. The pellets are fed by a conical screw mechanism into the boiler from the 100m³ (3531ft³) capacity storage unit. This project attracted a 30 per cent capital grant from the Bolzano provincial government.

Much larger pellet boilers than this exist to heat schools, hospitals and even whole towns. These will be looked at in more detail in Chapter 9, along with domestic, commercial, public and CHP systems that use pellets to generate electricity for local use or sale into the national grid.

What pellet technology offers the world

As a convenient, renewable and efficient fuel, pellets have much to offer the human race in the 21st century. With oil becoming scarcer, more expensive and more vulnerable in terms of global supply networks, a cheaper fuel that can be home-grown provides a very attractive alternative.

Climate change and pellets

The pellet industry has consciously developed to address a number of serious environmental, economic and political issues faced by the modern post-industrial and post-Cold War world. Perhaps the most important of all problems it helps mitigate is the changing climate. Our use of fossil fuels is accepted by scientists and politicians as the biggest single factor causing global warming, rising sea levels and fiercer, less predictable weather. Wood pellets, on the other hand, offer a practical and carbon-mitigating alternative.

Quality-controlled wood pellets are considered carbon-neutral (or almost carbon-neutral) by European Union countries and comply with the Kyoto Protocol position on air emissions, which goes so far as to credit pellet end-users with carbon savings. The carbon-neutrality argument is based on the fact that trees, the original source of raw materials for pellets, absorb as much carbon while growing as they give back to the atmosphere when burnt. There

Table 1.2 *Carbon dioxide emissions compared*

	Direct CO$_2$ emissions from combustion (kg/MWh)	Approx. life-cycle CO$_2$ emissions (including production) (kg/MWh)	Annual CO$_2$ emissions (kg) to heat a typical house (20,000kWh/yr)
Hard coal	345	484	9680
Oil	264	350	7000
Natural gas	185	270	5400
Liquid petroleum gas (LPG)	217	323	6460
Electricity (UK grid)	460	500	10,000
Wood pellets (10% moisture content)	349	15	300

Source: Dilwyn Jenkins; data from UK Biomass Energy Centre

are obviously variables to consider, such as the efficiency of the combustion unit and the fact that processing wood into pellets also has energy and carbon emission implications. The same is true of transporting pellets to end-users in large trucks. Sometimes, newly cut forestry wood is even mechanically dried before being processed into pellets, something that uses significant amounts of energy. According to a Department of Trade and Industry (DTI) study that looked at the life-cycle carbon dioxide emissions of various fuels (Elsayed et al, 2003), for every megawatt-hour (MWh) of energy utilized, wood pellets produce less than 5 per cent of the emissions of oil. When compared to natural gas, the same life-cycle analysts estimate that wood pellets produce only 5.5 per cent of the carbon dioxide emissions created by natural gas.

A study by the Salzburger Institute for Urbanization and Housing indicates that an average Austrian household switching from an oil to a pellet heating system would save up to 10 tons of carbon dioxide emissions every year. Their carbon-neutrality, plus the fact that wood pellets are a renewable resource, means that this technology often attracts capital grants and favourable policy instruments from regional, national and international authorities. Electricity produced from biomass (including wood pellets) already receives additional financial incentives in many countries. Heat from pellets is likely to follow suit.

As with all biomass heat provision, wood pellets are responsible for carbon dioxide emissions at the point of combustion. When burnt, levels of pellet emissions are similar to those of coal and almost double those of natural gas. Stove or boiler efficiency also affect overall efficiencies and emissions, so it's really important to burn pellets and other wood fuels as efficiently as possible, particularly during the next 50 years while global society struggles to mitigate and minimize the impact of our changing climate.

Recognized and respected as a renewable fuel, pellets also boast high levels of job creation and the tendency to keep income circulating in local economies. Jobs are created in forestry and woodland management, transport, pelletization, combustion equipment manufacture and sales. Since the mid 1990s, Austria has developed a pellet stove and boiler manufacturing industry worth over €100 million a year. In 2005 alone, Austrian pellet equipment manufacturers supplied 8874 wood pellet boilers and 3780 wood pellet heaters.

Security of energy supply

One of the big concerns among politicians and energy strategists today is the security of energy supply. Europe, for instance, depends increasingly on imported oil, gas and coal for space and process heat as well as electricity production. Imported oil and gas arrives by sea or through highly vulnerable overland pipelines sometimes coming from and going through countries with which political relations are arguably vulnerable and certainly unpredictable. With this in mind, governments are keen to ensure present and future energy supplies from resources that can be better controlled from home. It is argued that nuclear power can provide some security of energy supply, but so can energy sources such as wind, hydro, solar and geothermal, being based on domestic and renewable resources. Similarly, wood pellets and other biomass, such as wood chips, are also home-grown renewable energy sources.

At the time of writing, arguments rage around the concept of 'peak oil', i.e. the fact that one day we will have passed the point of maximum oil extraction and consumption. This is inevitable because fossil fuel reserves are finite. We may not, however, be able to identify the actual moment or even year of peak oil production until it is well passed. More oil and gas reserves are still being found all over the world. But when producers are sure that the fossil fuels they tap into are beginning to diminish, the volume of supply is likely to be minimized in order to maximize the value of remaining resources. At the same time, of course, the demand for fossil fuels globally is growing faster than ever, putting further upward pressure on the price of a barrel of oil. This scenario also relates to security of energy supply issues in the longer term. Some commentators, particularly from the environmental sector, claim that we have already passed the historic peak of oil production and that the age of fossil fuels has already ended. The oil industry tends to argue differently.

In a country such as Finland, with 80 per cent forest cover, wood energy, including pellets, is inevitably playing a major role in establishing a secure energy supply. Russia is now producing around 500,000 tons of pellets every year. North America produced 2 million tons in 2008, and, in the UK, with perhaps the best wind resource in Europe, wind farming is providing a significant alternative. With only around 10 per cent forest cover, however, wood can only provide part of the solution. Nevertheless, even in a small country such as the UK, more agricultural land could be dedicated to bioenergy production because it already imports pellets from Scandinavia, mainland Europe and even Canada. The potential for take-up even in a country with few woodlands could therefore expand with ease and relative elasticity. This can only be good, especially for colder countries where heat provision could become a major problem if winter gas and oil supplies are constricted, stopped or effectively reduced for any reason. With the major biomass resources geographically limited, it is possible to imagine the world market place dictating that a few major pellet-producing countries supply the fuel to all the other user-nations, creating new structures of international energy dependence.

Energy price trends

Nationally and internationally, energy prices appear more volatile and unpredictable year on year. Peak oil theory suggests that the price trend for this energy source, at least in the long term, is inevitably upwards. To date, this has been the global experience during the first decade of the 21st century. While the price of heating oil has almost doubled, the price of wood pellets has remained significantly cheaper. In November 2008, the price per kWh of oil was 4.5 pence in the UK; wood pellets (at £200 per ton) cost only 4.3 pence. Wood pellet prices did demonstrate a large price spike around the end of 2006 when demand for them began to outstrip supply following the massive surge in oil prices that began in 2005.

In 2008 the price of pellets rose again due to the global economic crisis, which saw a drop in manufacturing and thus in the availability of sawdust, a vital raw material for pellets. At the same time, oil prices went back down (see Figure 8.6, page 99).

Limitations to pellet technology

Pellet technology is proving good for providing heat at a wide variety of scales. There is no lower pellet stove size limit because pellet burners can be adapted to even the smallest of rooms. The requirement for an external flue even at a small scale is a disadvantage when compared to most electric heaters, but this is similar for gas and oil. The upper limit to pellet use is presently defined by Sweden, which consumed 1,670,000 tons of pellets in 2006, almost 60 per cent of which were used in large power stations. In Sweden, 2MW pellet boilers are considered medium-sized and Stockholm uses 200MW of biofuel-fired heat. The Hässelby plant, Sweden's largest user of wood pellets (consuming over 250,000 tons annually), provides 500GWh of heat every year from a mix of pellets and olive grains.

If an available supply of good-quality wood pellets can be assured at a reasonable price, then they offer a viable green option for heating anything from a room to a city suburb. Factories that produce their own wood chip waste or households surrounded by woodlands are situations where a chip or log boiler might make more sense environmentally, economically and in terms of fuel availability.

There are clearly physical limits to wood pellet technology due to the finite availability of the primary resource: woodlands, timber yards, joinery factories, municipal arboreal wastes and clean recycled waste woods. The environmental benefits of this fuel would be worse than negated, however, if forest (particularly primary rainforest) is unsustainably logged to provide material for wood pellet manufacture.

References

Canadian Wood Pellet Association (www.pellet.org)

Elsayed, M. A.; Matthews, R. and Mortimer N. D. (2003) *Carbon and Energy Balances for a Range of Biomass Options*. DTI, London IEA (www.iea.org)

IEA Bioenergy Task 40 (2007) *Global Wood Pellets Market and Industry*

Powys Energy Agency (2003) *Wood Fuel in Wales*

Rakos, Christian (2008) Director, proPellets, Austria (www.propellets.at)

Statistics Finland (www.stat.fi)

UK Biomass Energy Centre (www.biomassenergycentre.org.uk/portal/page?_pageid=75,163182&_dad=portal&_schema=PORTAL)

2
The Physics of Heat

We all have an intuitive understanding of heat and fire, which has long been important to human households for warmth and cooking. With the combustion of wood in small fires people were able to move energy from tree branches and direct it into cooking pots and the immediate space around the hearth. Our Stone Age ancestors knew perfectly well that by touching dry wood with a flame, raising its temperature, it could be combusted, and that once the wood had finished burning then only ash would remain. It must have been obvious even then that the wood was used up, its energy taken out as heat. Since that time, human ingenuity has developed technology and a whole civilization based around heat, power and the internal combustion engine.

Heat is energy in transit

The neat but more technical definition of heat is as 'energy in transit', always moving from a material of higher temperature to one of a lower temperature. As we know from heating our homes, this transfer of energy changes the balance of temperatures experienced by both materials. This definition was finally proven in the 1840s by James Prescott Joule. Energy, of course, cannot be made or destroyed, just transformed, or moved between different materials or states.

Combustion

Combustion is the process we use to induce temperature transformations, understood chemically as the super-fast oxidation of matter. Heat and sometimes light are consequences of combustion, which sucks oxygen in from the surrounding air to feed the chemical reaction. As it begins to burn, the temperature of the material rises. How much and how fast the temperature rises depends on the flow of air plus the quantity and qualities of the combusted material. Highly efficient combustion technology optimises the air-to-material ratio to achieve the desired combustion temperature while minimizing fuel input. Even Stone Age people knew how to blow on or fan a fire into flames. But they didn't know that the main by-products of combustion are gases, including carbon dioxide, carbon monoxide, methane and sulphur dioxide. Water vapour is another by-product of combustion, and there is usually ash left as material waste, although this is minimal when good-quality pellets are combusted (usually around 0.5 per cent ash content). When it comes to wood

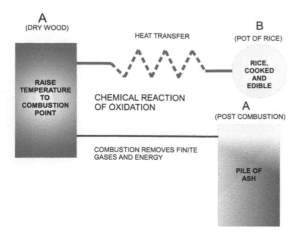

Figure 2.1 *Combustion for cooking*

Source: Teilo Jenkins

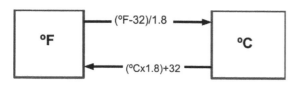

Figure 2.2 *Celsius to Fahrenheit conversion square*

Source: Chartered Institution of Building Services Engineers (CIBSE) *Domestic Heating – Design Guide,* Appendix D

Table 2.1 *Converting calories to joules and watts*

1 calorie	4.1868 joules
1 joule	1 watt-second
3,600,000 joules	1 kilowatt-hour

pellets, Hansen et al (2009) identified at least four consecutive phases of combustion:

- drying and evaporation of moisture (in early phase of combustion);
- gasification (also known as pyrolysis; the pellets emit gases as they are heated);
- combustion (the gases ignite above specific temperatures, their ignition points, visible when a smoking fire explodes into flames);
- coke burn-out (the burning-ember phase, leaving carbon particles as ash).

Measuring heat

Heat is measured in various ways. In everyday life, we tend to use degrees Celsius (°C) or degrees Fahrenheit (°F). Figure 2.2 shows how to convert between the two. In this book we use the metric system, so temperatures will be referred to as degrees Celsius.

It's also possible through the metric system to understand heat as an amount of energy that can be used to do work of some kind. This is expressed as kW or kilowatt-hours (kWh). Other standard metric units of heat and energy measurement are the calorie (or cal) and the joule. One calorie is equivalent to the heat required to raise the temperature of 1g of water from 14.5°C (58.1°F) to 15.5°C (59.9°F). One calorie is also equivalent to 4.1868 joules (J), another thermodynamic energy category.

A joule is sometimes described as roughly the amount of energy needed to lift an apple 1m (3 feet) off the ground or to heat a gram of dry air by 1°C (1.8°F). Within the accepted International System of Units (SI), one watt is a unit of power equal to one joule per second. One watt-hour (Wh) is 3600J and one kilowatt-hour (kWh) is equivalent to 3.6MJ. Or, to put it another way, 3,600,000J = 1kWh; that is, 1kW per hour of energy expended.

References

CIBSE (2007) *Domestic Heating – Design Guide,* CIBSE, London
Hansen, M. T., Rosentoft Jein, A., Hayes, S. and Bateman, P. (2009) *English Handbook for Wood Pellet Combustion,* National Energy Foundation, Milton Keynes

3
The Energy Source

Sources of pellet raw material

Sources of raw material for wood pellet manufacture are varied. In countries such as Sweden, the USA and Canada, which produce pellets on a large scale, feedstock for wood pellets comes mainly from the mountains of sawdust that would otherwise accumulate next to their numerous huge sawmills. In the UK, where there is less forest, some pellets are imported, although many are made from local sawmill or joinery waste material.

As long as it can eventually be reduced to dust or fine granules, any type of clean wood can be used to make pellets; coniferous, however, is the most commonly used around the world. Deciduous wood co-product is often carefully blended into the mix to optimize consistency of the final product in storage and during combustion. To date, sawmills continue to be the largest single provider of raw materials for wood pellets, followed by timber processors such as joinery yards making doors and windows. Municipal and other arboreal cuttings are a significant potential source for pellets. Freshly harvested whole trees, too, are being increasingly used to resolve bottlenecks and potential supply problems with sawmill and timber process by-product. Using arboreal cuttings or whole trees entails significant drying of the wood before pelletizing. Although it adds cost to the process, this could be covered, in theory, by using heat (sometimes freely available surplus heat) from a pellet CHP plant. Ideally, all tree bark is removed to avoid raising waste ash levels.

Recycled wood is often used, but it has to be 'clean' (see Chapter 7 for definitions and regulations), untreated and free of paint, heavy metals or other harmful additives. Any metal fastenings such as nails or screws must be entirely removed. In this form, such wood is more often used in large co-fired heat and power plants rather than small-scale stoves and boilers. Energy crops, or agri-fuels, from agricultural products – such as straw, hay, miscanthus or fast-growing willow – are an increasingly interesting option. Some early experiences with agri-fuels revealed that they burn less well and with more ash deposits than anticipated. This suggests that more product development, particularly in fuel conditioning, is needed before they are more widely available to the public. Even when agri-fuels are fully market-ready, their producers anticipate that the products will be used by large heat and power installations, probably in briquette form, rather than the domestic or public building sectors.

Quality of feedstock is vital for making good pellets. Before pressing in a pellet mill (see Figure 3.1), the woody matter must be rendered into dust or fine granules. This is generally achieved by shredding and/or pulverizing in a hammer

Figure 3.1 *Balcas pellet mills*

Source: Balcas

mill. If the feedstock's moisture content is more than 10–12 per cent, it will also need drying. Water (usually as steam) and sometimes vegetable oil, or starch, are added in a controlled fashion at the pressing stage to improve the consistency of the pellets (see Chapter 7 for more detail on pellet manufacture).

The finite availability of wood co-products, recycled wood and sustainably harvestable trees suggests that sourcing raw materials for pellets could become a major issue. If sawdust were to be imported from South America or Russia, where illegal and unsustainable logging is presently rife, such trade would require a very rigorous and transparent certification process to ensure that the source were sustainable.

Solar energy and carbon storage in biomass

According to the UN Food and Agriculture Organization, the amount of solar radiation (or insolation) striking the earth on an annual basis is equivalent to 178,000 terawatts (TW). This is between 10,000 and 15,000 times more than present global energy consumption and represents approximately 19MW per person (Burkhardt, 2007). Since not all of the spectrum of solar radiation is useable by plants, and because of other factors such as weather conditions and reflective plant surfaces, the overall efficiency of photosynthesis from the sun is limited to between 3 and 6 per cent. Furthermore, around 30 per cent of photosynthetic productivity occurs in the oceans, largely due to the activity of microscopic algae known as phytoplankton.

Biomass of any kind can be described as solar energy stored in organic matter. The largest single source of biomass is found in woods and forests, which are estimated to be 20 to 50 times greater than the biomass in agricultural land (Houghton, 2005). As they grow, trees photosynthesize the sun's energy, capturing and converting carbon dioxide from the surrounding air. To do this, plants use the chlorophyll in the leaves to capture the solar energy and turn it, along with water and carbon dioxide, into carbohydrates. These are transported around a tree to be used to make starches and cellulose for growth. At the same time, around a ton of oxygen is given off through leafy transpiration as a by-product for every ton of new woody growth produced. Surplus 'food' produced by a tree, over and above its growth requirements, is stored for the following year to develop leaves, flowers and seeds. Growing up from the tips of its branches and down through its roots, a tree also thickens around the trunk, forming annual growth rings.

Over time, the total biomass on the planet has changed considerably, not least because of human activity. Globally, forests are thought to have covered around 68 million km^2 (26 million square miles) during prehistoric times; nowadays, estimates suggest forest cover has been reduced by half to 34.1 million km^2 (Hutchinson Encyclopaedia, 2009). The main factors affecting the growth rates of forests, natural or plantation, include the spacing of trees, silvicultural treatment, thinning and pruning, site and climatic conditions. The latter factor appears to be kicking in as world climates change and the atmospheric balance of carbon dioxide rises. Plants need carbon dioxide to grow and, according to recent scientific research, trees in the tropics are presently growing faster (Adam, 2009). This is at least partly because of the greater availability of the gas and the evidence is that, compared to the 1960s, each hectare of intact African forest studied had trapped an extra 0.6 tons of carbon a year. Globally, this extra 'carbon sink' effect adds up to 4.8 billion tons of carbon dioxide removed – close to the total carbon dioxide emissions of the entire USA (Adam, 2009). In the USA's northwestern forests, however, trees are dying faster due to changing climatic conditions, which are warmer and dryer. Following the clear-cutting of a northern US hardwood forest, it takes about 40 years for the biomass levels to return to the pre-harvest values (Houghton, 2005). The future growth of forests and other biomass is increasingly difficult to predict at a time when our very climate is shifting in unpredictable ways and with as yet unknown consequences.

The carbon-neutrality of pellets made from clean wood depends on new biomass growth balancing (or outstripping) the biomass used for energy. Standing biomass is generally considered an excellent carbon sink and could be sustainably managed even for human combustion. Biomass planted specifically as energy crops is also likely to have a long-term role in the future of heat and power provision, but policy-makers will need to ensure that, at a time when land for growing food is scarce, any competition for arable land to plant energy biomass should be avoided. Energy crops are more appropriate for land unsuitable for agriculture.

The government of the US state of Oregon notes that biomass sources provide about 3 per cent of all energy consumed in the country. Biomass meets about 14 per cent of global energy demand and, in 2002, supplied around 47 per cent of all the renewable energy consumed in the USA, significantly more than any other renewable sector.

Despite these impressive figures, much biomass goes ungathered. With its traditionally large rainfall, the Pacific Northwest generates vast amounts of biomass each year, but competing uses plus the costs of collection and transportation are factors restricting the amount available for energy purposes. In Wales – where it also rains a lot– the average growth rate of conifers is estimated as yield class 12, which equates to an increment of 12m³ (423ft³) per hectare per year of stem wood or trunk timber at maximum mean annual increment (Potter, 2003).

In terms of climate change mitigation, the potential impact that biomass energy may have will depend on the scale of implementation and other life-cycle emissions associated with transportation and the efficiencies or inefficiencies of combustion.

Case study: Balcas Pellets

As one of the UK and Ireland's largest wood product suppliers, Balcas has access to vast quantities of its own sawdust and off-cut co-product, but also buys some raw material in from other sawmills. In recent years, it decided to start manufacturing wood pellets, at one stroke eliminating a waste issue and entering a rapidly growing new market sector. Furthermore, on one site Balcas has a 2.5MW CHP plant running off wood chip and providing heating and electricity for the site as well as selling electricity into the grid to serve around 10,000 homes. Balcas Timber expanded from a single site in 1988, when it acquired extra facilities, including three additional sawmills, a medium-density fibreboard processing plant and a pallet business.

Following a £15 million investment at the company's Enniskillen head office site, it has become one of the most advanced sawmills in Europe. It achieves

Figure 3.2 *Balcas* brites *wood pellet bags*

Source: Balcas

very high yields and line speeds (up to 120 linear metres per minute), with computer-controlled log positioning and saw setting, as well as optical vision systems creating 3D images of incoming logs. An automatic stacker handles 120 pieces of wood every minute and there's an intelligent automatic 100-bin sorter.

The company's production is more than 150,000m3 (5,297,200ft³) and the turnover is more than £70 million a year; it directly employs 700 people with a further 300 working on forest harvesting and haulage. The company operates from several sites, including sawmills in Northern Ireland, Scotland and Estonia, plus an architectural mouldings factory in Kildare, the Republic of Ireland. Overall, the company has an extensive product portfolio that includes construction timber, fencing products, internal mouldings, pallet and packaging products and wood pellets (branded by Balcas as 'brites'). Its sawmills and wood processing sites provide ample sawdust and off-cuts for the complementary pellet manufacturing process.

Balcas brites wood pellets are produced at two sites with separate wood pellet mills, one in Enniskillen producing 55,000 tons a year, the other, serving the UK mainland market, in Invergordon, Scotland, producing around 100,000 tons a year, enough to heat about 20,000 homes. Built on the site of an old smelter, this plant expects to offset around 17,000 tons of carbon emissions annually. The site will be self-sufficient in electricity and expects to export about 5MW of electricity (MWe) to the grid. The feedstock for its pellets comes from sawdust and other wood fibre.

In its first few years, Balcas has already built a strong, market serving more than 3600 homes and supplying in bulk to leisure centres, hotels, hospitals, schools, commercial offices, local authorities and prisons. To assist with its supply chain, a brites depot in Brentwood, Essex, serves southern England, while customers in Munster, the Republic of Ireland, receive their brites direct from a depot in Cork. The company's product is Forestry Stewardship Council (FSC)-stamped to demonstrate the sustainable source of its raw materials for the pellets.

Quantities of wood pellets consumed under normal conditions

When it comes to designing a new heating system (see Chapter 5), one of the factors you will probably wish to analyse is the predicted consumption of fuel. This will depend on the size of the system installed and, of course, your space and water heating requirements. The single most critical factor in most cases will be the amount of space to be heated. This will help determine the rated output of the heater or boiler. Once that has been decided, then the manufacturer's data will give details on fuel consumption at normal operating temperatures.

It's important to size a wood pellet system correctly. If it's undersized, then it will fail to meet full demand and will wear out faster. This will damage the heater or boiler. If, on the other hand, the system is oversized, it will be left to burn very slowly, using the wood pellets inefficiently and increasing waste products such as emissions, soot and ash.

Table 3.1 *Pellet consumption for typical uses*

System type/location	Space (m²)	Average system output (kW)	Annual pellet consumption (tons)
Heater/large living room	50	4–8	0.5–1.5
Boiler/typical house (space and water heating)	200	12–20	3–5
Boiler/typical school (space and water heating)	1000	120	30

The figures in Table 3.1 assume UK weather conditions plus high levels of insulation and draught-proofing. The US Department of Energy (DoE) has developed a free online downloadable spreadsheet to help compare fuels, including wood pellets, and heating system types, which can be used to calculate heating fuel requirements.

One rule of thumb used by bodies such as the British Columbia Sustainable Energy Association in Canada estimates that for every 10m² (107ft²) of living space to be heated, you'll need an input of between 1kW and 1.5kW of heat. These figures are a handy reckoner, but nevertheless only hypothetical and very generalized. In reality, wood pellet consumption will depend on a host of factors, including:

- internal and external average temperatures;
- months of winter heating needed;
- hot water consumption for kitchen, bathroom and other processes;
- the quality of wood pellets used;
- other heating inputs and internal dissipation (e.g. electric space or water heaters, other electrical appliances);
- any additional heat requirements (e.g. cookers or even swimming pools in the case of some schools).

To optimize fuel consumption, larger installations such as schools and hospitals are likely to make use of more than one boiler. The smallest of these boilers will usually be seen as a peak-load boiler, kicking in only when the demand is at its highest levels, or as a back-up in case of maintenance or even breakdown of the main boiler.

In conclusion, wood pellets are a highly versatile fuel source which provides a useful and valuable new market for by-products of the forestry and timber industries. They are carbon neutral, which means they can help many nations reduce overall carbon emissions. Pellets are also a relatively compact and fluid moving wood fuel, making them cheaper and easier to transport and deliver, and more convenient to use, than logs or wood chip. There are technical matters regarding combustion efficiency and fuel consumption but these are design issues.

References

Adam, D. (2009) 'Fifth of world carbon emissions soaked up by extra forest growth, scientists find', *The Guardian* online, 18 February, www.guardian.co.uk/environment/2009/feb/18/trees-tropics-climate-change

Balcas website, www.balcas.com/site/default.asp (accessed 19 November 2009)

Brites (Balcas product) website, www.brites.eu (accessed 19 November 2009)

British Columbia Sustainable Energy Association (www.bcsea.org)

Burkhardt, H. (2007) 'Physical Limits to Large-Scale Global Biomass Generation for Replacing Fossil Fuels – Insolation: the Physical Base of Green Energy', *Physics in Canada*, vol 63, no 3, pp113–115, www.scienceforpeace.ca/files/july07-burkhardt-proof1.pdf

Government of Oregon website, www.oregon.gov/ENERGY/RENEW/Biomass/BiomassHome.shtml#overvie (accessed 18 November 2009)

Hi-energy website, www.hi-energy.org.uk/balcas-invergordon-biomass-plant.htm

Houghton, R. A. (2005) 'Aboveground Forest Biomass and the Global Carbon Balance', *Global Change Biology*, Vol 11, pp945–958, www.whrc.org/resources/published_literature/pdf/HoughtonGCB.05.pdf

Hutchinson (Farlex) Encyclopaedia online, http://encyclopedia.farlex.com/Forest+structure (accessed 18 November 2009)

Miyamoto, K. (ed), (1997) *Agricultural Services Bulletin 128: Renewable biological systems for alternative sustainable energy production.* UN Food and Agriculture Organization, Rome Potter, P. (2003) *The Potential Sustainable Biomass Fuel Resource from Forest and Woodland Sources within the Dyfi Valley Eco-Partnership Area* TAPARES Project, Energy Agency for EC DGTREN, PowysPower Technology website, www.power-technology.com/projects/balcas*(accessed 19 November 2009)*

US Department of Energy website, www.eia.doe.gov/neic/experts/heatcalc.xl (accessed 19 November 2009)

4

Wood Pellet Stoves and Boilers

In this chapter, we will look in more detail at what distinguishes wood pellet stoves and boilers from other comparable heating appliances. The main operating features are described, as is the functioning of stoves and boilers themselves. The integration of other heating inputs, such as solar thermal, is also explained.

Wood pellet stoves – with and without back boilers

Wood pellet stoves such as those seen in Figure 4.1 are turned on and off at the press of a button or the flick of a switch. They are programmable, thermostatically controlled and provide efficient and environmentally responsible heat. The early prototypes for wood pellet stoves were developed first in Scandinavia and North America in the 1980s. Nowadays, a wide range of pellet stoves is manufactured, mainly targeted at markets in Europe and North America. Some are very smart-looking models, others are more simple basic metal boxes. Their main advantage over ordinary wood stoves is their convenience and cleanliness. Their main disadvantages include the higher cost of pellets and the need for electricity to drive their fans and sometimes their pellet-feed mechanisms.

One of the key benefits of pellet stoves is the nature of the fuel itself: clean, compressed and usually supplied in 10–25kg (22–55 pound) plastic bags, making it easy to load an indoor hopper and avoid the mess associated with buckets of coal or arms full of log wood. Pellet stoves can operate with high efficiencies even at relatively low heat outputs. High efficiencies, thermostatic controls and electric ignition mechanisms all add to their attraction. Style, functionality and price vary significantly, from highly glossy living room centrepieces to those little different visually from a basic stove used for burning wood logs. Pellet stoves circulate the hot air they produce using a fan, while most wood log stoves use the radiant heat of their hot metal casings. With pellet stoves, the casings tend to remain relatively cool.

Pellet stoves tend to be more expensive than wood log stoves of comparable output and, at least in the early days, they were often criticized for being noisier because they incorporated blowers or fans, one to direct the air for combustion and another to distribute the heated air into the surrounding room space. Another potential disadvantage is the pellet stove's reliance on electricity for

Figure 4.1 *Rika wood pellet stove*

Source: Rika

pellet feeder and fans. Without a back-up power supply, the stove will stop working in the event of power cuts, which generally happen in deep winter conditions. Pellet dust (basically, very fine sawdust or broken lumps of pellet) is considered an issue by some users. In hand-loaded pellet stoves, for instance, dust can accumulate behind and around the base simply because you have to lift the bag to feed the hopper. The higher quality the pellet, the less often this will happen. It's also possible to load pellets from a bucket with a spout (or a watering can with its spout cut short to improve flow conditions), which will help to avoid any dust problem. Correct flue installation is very important even for small pellet stoves, and in some circumstances a separate safety flue is required.

The pellet burner and various automatic options distinguish wood pellet stoves from the generally simpler wood log stoves. Both stove types have to handle the intake of cold air, the combustion and the safe extraction of smoke, but the options for automatic controls, motorized fuel-feed mechanisms and powered air circulation are typically much more common for wood pellet stoves. This is the main reason they are more expensive than wood log burners.

Over the past 20 years, the pioneering nations of Sweden and Canada have been joined by some other countries with their own large forestry industries. Where this coincides with progressive energy policies, as in Austria and Denmark, these countries have created strong wood pellet manufacturing bases that have helped to expand the market for pellet stoves, pellet boilers and some with integrated solar thermal input. What they all claim to offer is convenience, efficiency and automation.

Generally speaking, wood pellet stoves are used for single-room space heating. Occasionally, however, they may be needed to provide heat to other areas or spaces of a domestic, commercial or public building. The most efficient way to do this is to use a stove with back boiler. Another alternative is to allow heat to circulate beyond the room where the stove is installed by radiation or the circulation of warm air. Virtually all pellet stoves have some level of automation and require an electrical input for ignition, a small motorized fuel feed (or target temperature) mechanism and one or two motorized fans for combustion, flue gas and warm air flows. Most stoves have a manual damper knob to control air flow, to maximize combustion efficiency and optimize heat output.

For most potential users, researching wood pellet stoves on the internet and discussing options with two or three local pellet stove dealers is probably the most practical way to begin identifying which stove suits best. Ultimately, though, selecting the right type and potential kilowatt output of pellet stove depends on a few factors, relating primarily to end-user requirements and budget:

- price and available budget;
- space heating requirements (see Chapter 5 for details on calculating this);
- physical constraints relating to stove and flue location (in relation to the way that the space is used e.g. where people sit in the room);
- stove types available through local dealers (who should be able to advise on technical specification, supply the equipment and install the stove to the manufacturer's standards, with a warranty and affordable, guaranteed technical support);
- availability of complete operator's manual for the stove (ideally in your first language);
- stylistic taste (arguably the most subjective factor, but nevertheless very important).

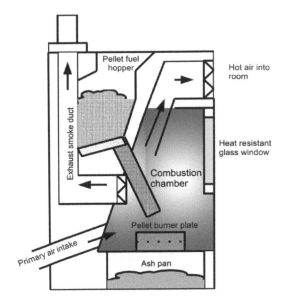

Figure 4.2 *Diagram of wood pellet stove operation*

Source: Teilo Jenkins

Main operational features of a wood pellet stove

This section describes in a little more detail most of the operational features of a pellet stove (see Figure 4.2). Many of these features are also associated with most wood pellet boilers, although such systems also incorporate other features including more sophisticated controls.

Pellet fuel hopper

Most wood pellet stoves have fuel hoppers (containers where pellets are stored ready for feeding into the burner) integrated into the upper section of the stove. These are filled by hand, usually pouring the pellets from a plastic bag into the lid of the stove, which is often used as the hatch door for the integrated pellet hopper. The higher-end hoppers allow for a slight further drying of the pellets and are usually designed to use gravity to drop-feed pellets into the pellet burner via a screw auger mechanism. Hoppers tend to be large enough to hold fuel for one or two days, roughly 10–25kg (22–55 pounds) of pellets. Some stoves' hoppers will take an extra 5kg (11 pounds).

Pellet feed tube

The pellet feed tube allows pellets to flow from the bottom of the hopper down towards the pellet burner. This is usually regulated by an inclined screw auger feed mechanism. In the event of there being an external pellet store or hopper, a separate auger is required to bring the pellets into the boiler. One of the beauties of pellets is that they can be fed one by one into the burner, optimizing heat output efficiencies and fuel economy.

Pellet burner

Pellet burners vary in size, but they are all surprisingly small. In Scandinavia, they can be found at their most compact and are available as independent units for retrofitting into some appropriate oil or solid fuel stoves or boilers. Their basic design is a burning plate for combustion of small quantities of pellets, plus a direct air blower to assist combustion (see below). Pellets are either dropped into the burning plate (that is, overfed), fed horizontally or, in many boiler systems, underfed. With horizontal and underfeed methods, the pellets are delivered by a stoker screw, the auger, direct into the burner. Overfed burners drop the pellets into the burner and this offers added fire protection (against burn-back – see Chapter 5) by separating the hopper physically from the burner. Dropping them allows for minutely accurate delivery of pellets, but disturbs the combustion process, causing more ash and particles than equivalent stoker-fed burners. Once the pellets are ignited (see below), they burn on the plate, typically with a rising yellow flame. The heat at the burner plate is usually maintained around 150°C (300°F), enough to send the volatile gases up into the combustion chamber. Most burners will require monthly cleaning.

Combustion chamber

Combustion chambers are required to achieve good combustion quality. For this a space (or chamber) is designed to optimize the three Ts: effective turbulence (for the gases to mix well), the right temperature and the correct length of time to complete the process. The combustion chamber may extend up to about 45cm (17 inches) directly above the pellet burner. The shape varies significantly from a simple, relatively inefficient box, to cylindrical chambers connected to an array of cold air inlets, hot air ducts and, in the case of stoves with back boilers, sophisticated heat exchangers. Target temperature ranges in the combustion chamber are generally between 800°C (1472°F) and 1000°C (1832°F). Temperatures significantly below this in the combustion chamber may cause unburned hydrocarbons to escape to the flue and into the outside air. This reduces efficiencies, can cause sooting-up and damage both the boiler and flue, and can also lead to unwanted air pollution. Combustion chambers may be made from stainless steel, silicon carbide or firebricks. Silicon carbide is the most expensive and the most resistant to slagging and corrosion. Stainless steel is fairly resistant. Firebricks are more expensive and have a tendency to slag up.

Ignition devices

Once it is sized (see Chapter 5), selected and installed (see Chapter 6), it will be time to light the stove. Most stoves on the market come supplied with a simple automated electronic ignition device. Some of the more basic stoves are designed for manual lighting, but the majority offer electric ignition.

Manual lighting
Stoves that require manual lighting are usually very simple to use. First, a small handful or cup of wood pellet fuel is placed directly into the burner or firebox. The use of a firelighter product (such as alcohol- or paraffin-based gel or

sticks), lit with matches or a gas lighter, is the most common method of lighting non-automatic start wood pellet stoves; but small amounts of well-dried paper or cardboard also work well and have the advantages of economy and leaving less chemical residue.

Electric ignition

Designed to provide that added element of automation, this type of ignition device is essentially a small electric element located in the fire basket and operated by a simple on/off switch, usually located on the stove's outer casing along with other controls. Stoves with automated electronic ignition are lit with the firebox door closed and the air blower switched on. The door on manually lit stoves should be closed as soon as possible after ignition and the air blower switched on immediately. Along with the rate of fuel feed, both the air flow (mainly a consequence of flue design) will affect both ease of ignition and efficiency of stove combustion.

Ash pan

A major difference between more conventional (electric, gas or oil) and pellet or wood heaters is that the latter two require human intervention in the form of regular ash removal. The amount of ash should be much less in a pellet stove than a standard wood burner because of the dry, compact nature of the pellets themselves and the efficiency of most pellet combustion technologies. Almost always located directly underneath the pellet burner, ash pans in most pellet stoves and boilers should be emptied every two to four weeks. Some of the more sophisticated boilers incorporate automatic daily ash compression into their ash pan, allowing for greater time spans between ash removals.

Figure 4.3 *Boiler ash chamber with pan and compacting device*

Source: ÖkoFen

Primary air intake

The primary air intake pulls air from outside the stove (or boiler) and directs it straight into the pellet burner. On ignition and warming of the pellets, this air is essential to raise the combustion temperatures sufficiently for the gases to be released from the heated and smouldering pellets. Some appliances depend on natural air flows for the primary air intake, while more sophisticated models tend to use a motorized fan to boost and control the lighting and combustion process. Air staging permits complete combustion and keeps emissions low. Separating the chamber into primary and secondary combustion zones through the furnace's geometrical design and with separate air feeds improves the secondary oxidization of flue gases.

Hot air escape tube

The hot air escape tube (or tubes) takes heated air from the combustion chamber either horizontally or vertically. For stoves with back boilers, these tubes sometimes pass over or through heat exchangers. Eventually, the hot air tube rises to a grille on the upper and outer levels of the stove. From here, it is usually expelled by a hot air circulating fan or blower.

Hot air circulating fan/blower

Hot air circulating blower units are usually small fans that require 150W, or even less power, to operate. Their function is to distribute the hot air into the available room space.

Exhaust smoke duct

The exhaust smoke duct takes the spent (and any unspent or partially burned) gases and other combustion emissions up into the stove (or boiler) flue and chimney system for eventual removal into the outside air. Pellet stoves should be considered along with other solid fuels in terms of their flue requirements. This is a very important element for the safety of any pellet heating system and is considered in more detail in Chapter 5.

Smoke extractor fan

A smoke extractor fan removes the exhaust fumes away from the combustion area and up into the exhaust smoke duct. In some stoves (and boilers) this fan is operated by the same motor as the primary air intake fan.

Heat-resistant glass window

As with many modern wood burners, pellet stoves can serve as a stylish centrepiece in a living room. In this common situation, users can enjoy the sight of the flaming combustion by selecting a stove with an ample glass window on the front casing. Most pellet stoves offer this option. All use heat-resistant glass.

Automation and controls

Various levels of automation and sophistication in controls are offered by pellet stoves, but the most common are an on/off button plus some kind of thermostatic control switch to regulate heat output (by controlling pellet fuel consumption and air intake using factory-programmed settings and in-built sensors for constant self-diagnosis). No stoves on the market automate absolutely everything; however, only gas and electric heating technology need less user input. Pellet stoves usually just require the operator to undertake regular hopper loading, ash removal and occasional cleaning. Some boilers have automatic ash

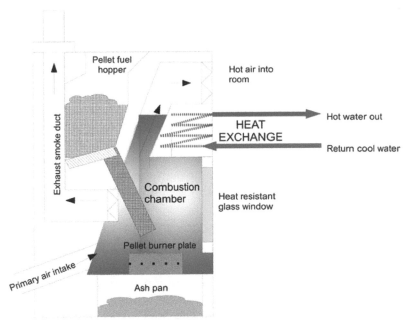

Figure 4.4 *Diagram of wood pellet stove with back boiler*

compression, which lengthens intervals between de-ashing, and a few boilers offer a screw auger to convey ash from the pan to a deposit, making it possible to empty ash just once a year. Software that measures temperatures, permits timer control, monitors flue gas concentrations and coordinates safety devices is found in all the more expensive stoves and most pellet boiler systems. As mentioned above, most stoves now use electric ignition devices.

Back boiler

Where a back boiler is incorporated into the pellet stove, the hot air circulating fan will typically have between 20 per cent and 60 per cent of the heat available for distribution; the rest will be taken up by the back boiler unit and dispersed through a wet (water-based) radiator central heating system, in underfloor piping and/or for domestic hot water. The size of the heat exchanger in a pellet stove determines whether or not a separate safety flue and an open expansion tank are required (see Chapter 5).

Pellet central heating boilers and how a conventional central heating system works

Introduction to pellet boiler systems

Putting aside the issues and technologies of wood pellet supply, storage and boiler fuel-feed mechanisms, a good wood pellet boiler system is not so different to most other modern heating systems. Essentially, there's a boiler that heats water and circulates this around the property in pipes that give out the heat via radiators or something similar such as underfloor heating coils. Hot water is stored temporarily in some kind of cylinder. To move heat around the system, the hot water is usually pumped and there are clever user-friendly controls to adjust the temperature of the boiler water, radiators and individual rooms.

Like oil and gas boilers, wood pellet boiler systems are designed to function efficiently while meeting heat loads that may vary significantly even over short cycles or time periods. They can deliver hot water quickly, often without any need for a buffer or accumulation tank. Log boiler systems, on the other hand, are unable to modulate and cycle effectively and will normally incorporate a buffer tank to hold the hot water until it is needed.

Before trying to understand how a wood pellet boiler system actually works, it's probably good to get to grips with how conventional heating systems operate.

How a conventional central heating system works

Not everyone understands how their central heating system works. Even having read the boiler instruction manual and the literature that comes with the programmer, pump or thermostatic controls, the overall functioning of an integrated central heating system can still be a bit of a mystery to many users. Understanding how a conventional heating system works is also made more difficult by the various types of system being used.

Practices vary between countries and regions as well as between the different fuel types, but generally speaking, all heating systems were traditionally open-vented and operated at low pressures. Increasingly, with improved plumbing technology, the world is moving over to higher-pressure sealed heating systems. Whether high- or low-pressure, most central heating systems are designed to distribute heat around a property via the medium of water heated in a boiler of some kind. The hot water is distributed through a network of pipes and radiators (or similar devices such as underfloor or skirting board heating systems). Even the most basic systems will require a boiler to generate heat for space and hot water needs. Virtually all systems will need a pump and a diverter valve (or mid-position valve) to circulate the hot water. Radiators or similar are required to deliver heat efficiently to the property's internal spaces. As well as these, most systems are incomplete without some kind of cylinder to act as a temporary hot water store.

Boiler systems with cylinders keep the boiler and radiator hot water physically separated from the mass of water in the hot water cylinder. This allows normal mains water to be used in the cylinder, so water from hot taps in kitchens and bathrooms is clean and doesn't circulate in the boiler or radiators. It also permits the addition of corrosion inhibitor fluid in the sealed boiler/radiator water system. The two systems (boiler/radiator and hot water cylinder) are kept apart by a coiled pipe inside the cylinder that is plumbed into the boiler. Hot water is pumped into the cylinder from the boiler but it circulates in the enclosed coiled pipes, so doesn't ever have to mix with the cylinder water. Instead, the water is being heated indirectly from the boiler through contact with hotter coiled pipes in the cooler cylinder. For this to work properly, the thermostat in the cylinder must be set at a lower temperature than the boiler thermostat, which determines the actual heat of the hot water coming from the boiler into the radiators and cylinder. If they are set at the same temperature, or if the boiler temperature is set lower than the cylinder's, then the cylinder thermostat will be on constantly, defeating the efficiency objectives of having a thermostatic control for the cylinder.

There are two basic types of central heating systems: one is fully pumped, the other uses convection for maintaining the hot water cylinder (which must be located above the boiler). A system using convection will still normally need a pump to transport water to the radiators (unless the boiler is on the floor below space heating needs, in which case it could potentially be an entirely convective system). Convection systems suffer from inefficiencies. They need the boiler to be fully fired-up to work and tend to be installed in systems that are in constant use over the winter. They provide hot water to the cylinder through unpumped convection flows, with a water temperature determined entirely by the temperature of the boiler. In this kind of system, room thermostats can still be used, but their function is relatively crude, simply turning the radiator pump on or off and thereby transporting hot water to the radiators or preventing it. For the purposes of this section, we assume that boiler systems are pumped.

Most modern central heating systems – such as oil, gas or good wood pellet boilers – incorporate as standard sophisticated system and appliance controls. Today, these sometimes include external broadband links to the programmer and internal wireless connections to room thermostats. Once up and running, a boiler can be switched on and off according to demand, either by a room thermostat or the hot water cylinder thermostat. The boiler itself has an on/off switch and an ignition device. There will be an electronic programmer that can control boiler functions by applying operator-programmable stop and start times for the various functions (primarily space heating and hot water provision). These days, most central heating systems include individual room thermostats, making it possible for users to enjoy different temperatures in different spaces. Generally speaking, a room where people sit and do little may need a higher temperature than a workshop or kitchen space. Bedrooms, too, can have lower temperatures since we wrap up in bed for most of the time spent there. Radiators themselves are also commonly fitted with control valves that allow the user to prioritize radiators in certain rooms over others in the system.

Valves are needed in the system to help divert the hot water flow by giving priority to either the space heating by the radiators or the hot water requirements of the cylinder and taps. Both the room and cylinder thermostats will let the boiler, and an electronically operated diverter (or mid-position) valve, know when they need more hot water to do their job. The valve will switch the flow from one to the other depending on demand, but if both the radiators and the cylinder need hot water simultaneously, then it will automatically assume a mid position, allowing both flows at the same time.

A central heating pump overrun feature is common in modern boilers, which makes the pump keep running and moves water out of the boiler and around the radiators and/or cylinder for a little while after the boiler combustion switches off. This protection device prolongs the life of the heat exchanger by preventing any static water remaining in the boiler and potentially boiling as a result of any residual heat in the minutes following boiler switch-off.

Main heating system types used today

Open-vented and fully pumped

Fed by cold water storage header tanks, open-vented systems operate at low (atmospheric) pressure because they have feed (or expansion) tanks that are

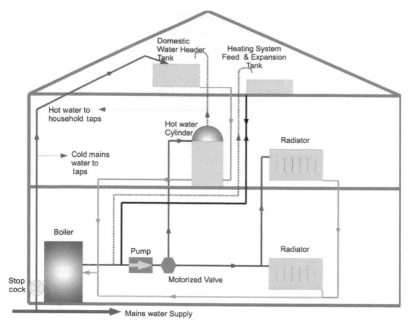

Figure 4.5 *Conventional open-vented system*

open to the air. Feed tanks are located above the rest of the system and are controlled by a ball valve allowing water lost by evaporation or leakage to be replaced automatically. These systems also incorporate a boiler, controls and a hot water cylinder. Feed (or in this case expansion) tanks also compensate for the expansion of hot water; they do so via a vent pipe that is connected from the pipework near the boiler (usually close to the feed pipe) and which empties into the tank. There are significant heat losses from the cylinder and also from the long pipe runs usually necessary between the boiler and cylinder.

Sealed system

A sealed system uses a sealed expansion vessel with sufficient cushion (acceptance volume) to accommodate the volume change when the boiler heats the system water from 10°C (50°F) to 110°C (230°F). These systems generally operate at relatively high pressure (around 1 bar above atmospheric pressure). To achieve this pressure, a mains water system filling loop is needed. Pressure can be read off an incorporated pressure gauge; during normal operations, pressure should be maintained without operator input, but it will lower if radiators are bled or the system leaks. Safety valve and venting are also required for a sealed system. The closer to the pump inlet that the expansion vessel is located, the easier it is to ensure positive pressures throughout the system during operation.

Compared to open-vented systems, certain economies are associated with well-designed sealed systems: flow water temperature can be higher, so fewer radiators (or other heat emitters) need to be used. Like conventional boiler systems, system boilers depend on stored hot water in a cylinder. Many of the individual components of the system boiler system are built in the factory, so pipe runs are often fewer and installations simpler, faster and less expensive than conventional systems. They are also economical to operate and fast to react. They don't need a feed or expansion tank or a vent pipe and the risk of freezing pipes is minimized. Water is supplied manually using a filling loop and an expansion vessel accomodates the rise in pressure during heating of the system water. Sealed systems can be vented low-pressure or unvented high-pressure; both need a hot water cylinder. But unvented system boilers incorporate a pressure relief valve and a pressure gauge, making a feed and expansion tank unnecessary.

Combination (combi)

Combination (combi) boilers provide an instant water heating solution with no need to store heated water. Apart from this, they are similar to conventional open-vented systems. Combis save capital and labour costs during installation and demand less internal space for the boiler system itself, primarily because of the lack of a hot water cylinder.

Combi boilers also operate without a header tank or cold water storage cistern in the loft. Water pressure even at the hot taps will be guaranteed to be at mains pressure, unlike some other systems, but hot water to more than one tap in the house at the same time can be problematic.

Condensing boilers

Condensing boilers operate with low greenhouse gas emissions and high efficiencies (86 per cent or even higher) because of their use of a larger or a second heat exchanger to extract additional heat from normally unburned gases. Condensing boilers can run on oil, gas or LPG.

Main features of a pellet boiler central heating system

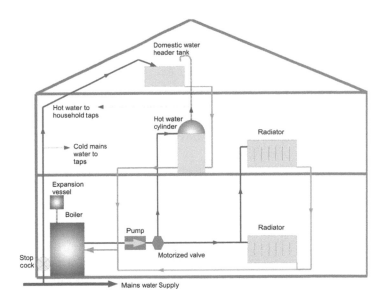

Figure 4.6 *Sealed system*

Source: Teilo Jenkins

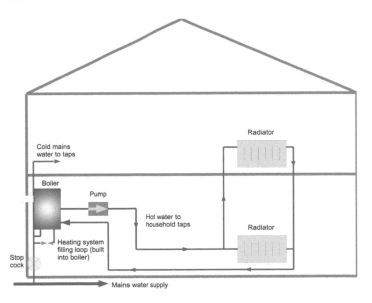

Figure 4.7 Typical combi system

Source: Teilo Jenkins

The main operational features of a wood pellet boiler are pretty much the same as described above for a wood pellet stove:

- pellet feed tube;
- pellet burner;
- combustion chamber;

- electronic ignition;
- ash pan;
- primary air intake;
- heat exchanger (round the hot air escape tubes);
- hot air circulating fan/blower;
- exhaust smoke duct;
- smoke extractor fan;
- automation and controls.

Additionally, a pellet boiler system will have:

- internal heat exchangers;
- heat distribution pipes;
- pumps;
- more sophisticated controls, sensors and programmes;
- a hot water cylinder (or possibly an accumulator or buffer tank);
- pellet storage room or tank;
- automated pellet-feed mechanism (usually an auger) to take the pellets from the store directly into the boiler;
- significantly larger footprint on the floor.

Quality wood pellet boiler design

In the past 20 years, Austria has earned itself a reputation as the bastion of top-quality wood pellet boiler design. Many Austrian pellet stoves and boilers are considered the Rolls-Royces of their type. Among the biggest manufacturers in the country are Köb, ÖkoFen, KWB and Fröling. One of the world's leading manufacturers of heating and renewable energy systems is the Viessmann Group, established in Germany in 1917. It now employs over 8000 staff internationally and has a turnover of around €1.4 billion. It has 12 manufacturing facilities in Germany, France, Canada, Poland, Austria and China, plus sales and distribution facilities in Germany and 35 other countries. Coming into the pellet business around 20 years ago, ÖkoFen specializes in manufacturing wood pellet boilers and associated component parts and is a great example of best practice in this relatively new industry. In this company's view, to be sure that a pellet boiler system works it's essential to have all the components completely compatible, so it now produces all the parts of the system to its own specifications. It produces low-mass, fast-reacting boilers.

Unlike many Scandinavian systems, which tend to have simpler controls and be based around horizontal air tubes, most Austrian boilers use vertical tubes. There has been a tradition in Scandinavia of adapting oil boilers by removing the original oil burner and inserting a pellet burner. This is not ideal for facilitating high levels of boiler modulation and cycling (see below for description of these terms) required for the UK and some parts of the USA, where the winters are relatively mild. This is very different to Scandinavia, where the winters are long and often bitterly cold; in temperate zones, a boiler that can modulate and turn on or off according to minute-by-minute demand, will prove much more cost-effective and efficient.

Boiler design

Austrian boilers tend to use vertical fire tubes with automatic cleaning devices (metal coils) built into the ducts. Sometimes these coils are designed for regular manual cleaning, but increasingly they offer full automation and are operated by a small motor.

Underfed burners are common as a method of delivering pellets in smaller boilers. Temperatures at the burner plate are ideally maintained around 150°C (300°F) during combustion. This avoids accumulation of solid ash and clinker that can happen at temperatures between 200°C (390°F) and 250°C (480°F). Some companies, such as KWB, use a rotating ash scraper on the burner plate to permit higher burn temperatures.

While 150°C (300°F) is an ideal temperature for the first stage of gasification, it is the secondary combustion that produces most of the heat in a good pellet boiler, so manufacturers aim for combustion at over 500°C (930°F) and preferably between 800°C (1470°F) and 900°C (1650°F) to ensure maximum gasification. Well-designed boilers will then recirculate some of the partially burned flue gases, which tend to be more sluggish (slower moving). They can be drawn back out of the flue as they pass the heat exchanger, then sent back into the combustion chamber as a fuel for what is basically a form of tertiary combustion but which also cools the flame and helps reduce nitrogen oxide (NOx) formation. Air ducts often require regular cleaning; some boilers do this automatically with motor-controlled springs that move up and down to clear any deposits on the inner walls.

Figure 4.8 *Close-up of automatic cleaning springs in an ÖkoFen PELLEMATIC boiler*

Figure 4.9 *Looking down into an ÖkoFen PELLEMATIC combustion chamber, with pellet burner at centre bottom and cleaning springs in a ring*

Boiler modulation

The ability of quality wood pellet boilers to modulate responsively is one of their main advantages over other wood heating systems, particularly wood log, but also to a lesser extent wood chip. As well as offering a much more fluid fuel product that can be delivered without blockages and in small amounts, pellets are much drier than wood chips and can be fed in very small quantities into the burner where the action of modulation is principally located. The pellet burner itself controls primary combustion air and the pellet fuel feed which are important for modulation requirements. Modulating controls monitor the system's hot water temperatures in relation to demand at any given moment. By knowing the demand and the existing provision of hot water, the boiler controls can determine how much, if any, wood pellet fuel to feed up to the burner plate. Wood pellets are small, compact and energy-dense, and they can be fed by a controlled auger virtually one by one into the small burner plate. Combustion in pellet burners can be rekindled just by increasing air flow for up to an hour after turndown, so this continuous and rapid matching of fuel and air input to temperature demand can help improve overall boiler efficiency enormously. (Turndown is the moment at which the boiler stops asking for more fuel, i.e. it has been manually or automatically turned down or temporarily 'off', in that there is no longer any request for more pellet fuel to burn in the combustion unit.) Having what is known as wide boiler turndown ratios, most wood pellet boilers can turn up the heat very quickly just by adding more fuel and/or air and, similarly, turn it right down by instantaneously switching these off.

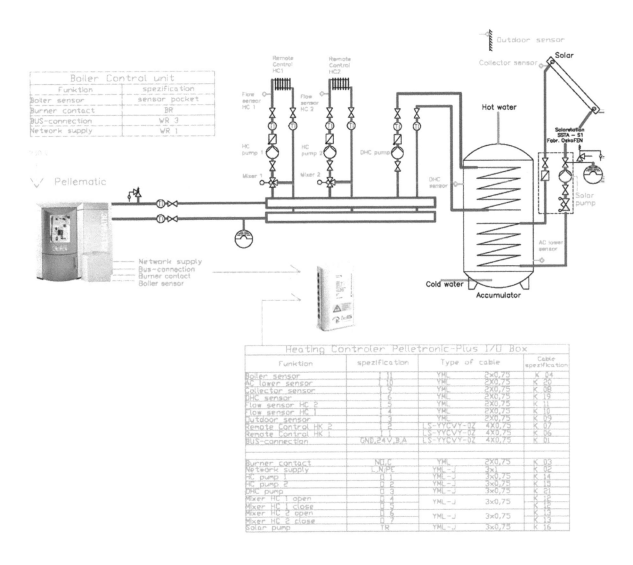

Figure 4.10 Typical hydraulic design

Source: ÖkoFen

Connecting diagram 1

1 Pellet boiler, 1 Accumulator - Pellaqua inc. domestic hot water system, 2 Heating circuits, 1 Solar

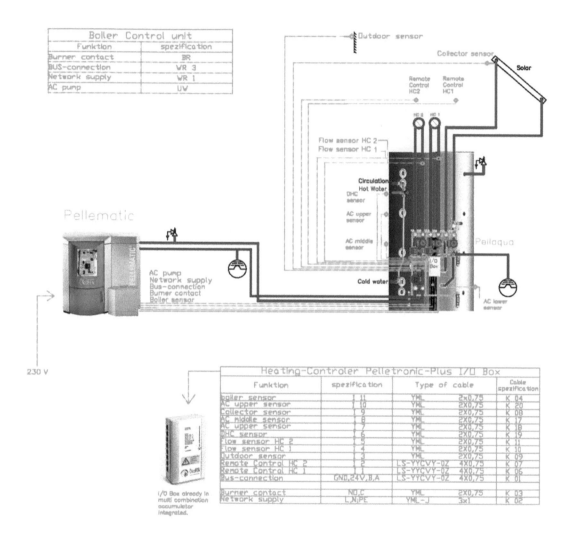

Figure 4.11 *Hydraulic design incorporating buffer tank*

Source: ÖkoFen

Boiler cycling

Boilers are often incorrectly sized, usually being oversized for their predicted heat loads. This may be because whoever designed the boiler wanted it to be able to cope with the worst of winter scenarios to ensure it was never insufficient for the task in hand. In temperate countries, where winters can phase into peak periods of cold, there can be weeks or months where the boiler is turning itself off and on if it is oversized. A boiler system will lose heat during off periods and it takes time to reheat it. One definition of cycle efficiency is annual average efficiency. More specifically, this is the ratio of annual heat output to the annual fuel energy input. Low ratios are usually due to excessive cycling. Generally speaking, the annual efficiency of wood pellet boilers is relatively high because they combine smart programs for boiler optimization control, automated pumping stations and, sometimes, the use of a hot water buffer tank.

The hydraulic system

Pellet boiler hydraulic design obviously varies, not just from make to make but also from installer to installer.

The question of when to have a buffer tank and when not to have one is a commonly asked one. It adds capital cost and requires additional floor space, but it can make a system more efficient. Boilers that modulate and can hold their embers well for an hour or so don't generally need a buffer tank in the system. But if you want stabilized combustion, this is easier to achieve with a buffer tank in the system. For instance, a building with a very low heat load or highly variable heat and hot water requirements can usually benefit from a buffer tank to help reduce modulation demands. Having a buffer in these circumstances allows for an increase in cycle length, which improves the ratio of annual efficiency (described in the section above on boiler cycling).

Sometimes buffer tanks are incorporated into boiler systems specifically to permit rapid combustion. In these circumstances, hot water distribution will come primarily from the buffer as demanded by radiators, underfloor heating or bath. Sophisticated boiler systems can also mix hot water direct from the boiler with preheated water from the buffer tank, or cold water, to achieve the correct temperature in the radiators or to the taps. Wherever possible, boiler systems should aim to avoid having hot water sitting unused in the tank for very long.

Most of Europe has been using sealed systems for a generation or two. Slightly behind the times, UK regulations demand that 'solid fuel heating systems' should not be sealed (see Chapter 5 in this volume and the HETAS Guide, www.hetas.co.uk/public/hetas_guide.html).

Boiler system controls

Interface with a wood pellet boiler is designed to be as simple and user-friendly as with other modern central heating systems. This automatic operation depends on controlling the load and combustion together. Load controls over fuel and air supply are governed by feed-water temperature. Combustion control is determined by a combination of the instantaneous heat demanded,

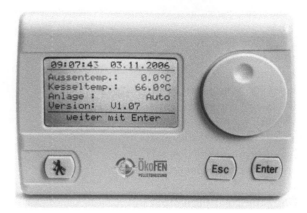

Figure 4.12 *Compact heating circuit controls*

Source: ÖkoFen

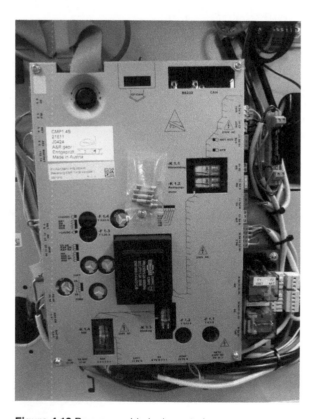

Figure 4.13 Programmable logic control

Source: ÖkoFen

the rate of fuel input and the sensors. The sensors register the oxygen and/or carbon monoxide concentrations in the flue gas and help the boiler's programmed controls decide on the appropriate primary and secondary air intake rate. Sensors with a proven record (for example, Lambda) are available on the market (and those for oxygen are cheaper). Figure 4.12 illustrates a standard ÖkoFen pellet boiler's microprocessor-based compact heating circuit controls, with digital text-capable screen and with the capacity for intelligent controls and timers, not only for the boiler itself but also for the cylinder or buffer tank, pumping stations and room thermostats and over auxiliary heating systems such as a solar thermal input. Each of the system's heating and hot water circuits can be adjusted for temperature and timings.

The heating circuit controls illustrated in Figure 4.12 are equipped with only one dial and two buttons, so the entire heating system can be regulated simply by pressing and turning. Menus are straightforward, with a simple text display. Many boilers can be controlled from other rooms in the house with digital or analogue remote control. Remote access to boiler via mobile phone or broadband is also possible.

Start-up and shutdown cycles are built into most boilers' programs. Lambda electronic control systems are commonly used for wood pellet boilers. However, the control panel, circuits and underlying programming in some systems is typically designed in-house according to the company's own programmable logic control (PLC; see Figure 4.13) system. Company programs are worked out by a combination of complex calculations and decades of experience.

Available system/configuration types

There are a great many types of wood pellet stoves and boilers on the market. Stoves are limited in their heat output and are most commonly used simply to heat a room. Several companies produce larger stoves (12–18kW) that can heat multiple rooms or, at a push, a small house, via a sealed or open-vented system with a wet heating and/or hot water distribution circuit. When it comes to wood pellet boilers, the range of options is much greater and they can cater for the heating requirements of anything from a small house to a school, hospital or village.

Most boiler system configurations are possible with wood pellet technology. It's also highly compatible with solar water heating (dealt with in more detail below). Wood pellet boilers can be used in conjunction with other boilers, either as the main boiler or a back-up to meet peak and low loads more efficiently.

The three scenarios shown in Figures 4.14 to 4.16 are basic designs commonly associated with wood pellet boiler systems (assuming UK weather conditions).

Stoves

Stove without back boiler

Clearly the simplest wood pellet heating device available, stoves require a flue and electricity point but not plumbing. Essentially, they consist of:

- fuel hopper;
- burner and combustion chamber;
- fan-assisted warm air circulation.

Stove with back boiler

Stoves with back boilers offer more system options and require plumbing into the domestic hot water system (via the cylinder) and a compatible wet heat distribution circuit to either radiators or underfloor piping coils. The main elements of this system include:

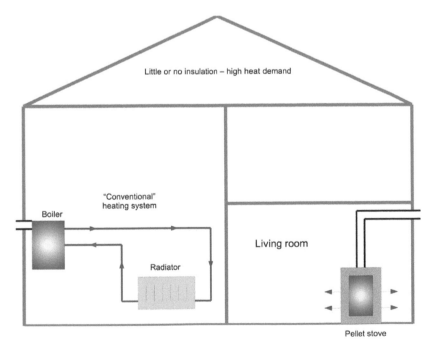

Figure 4.14 *Wood pellet stove in typical older house augmenting space heating*

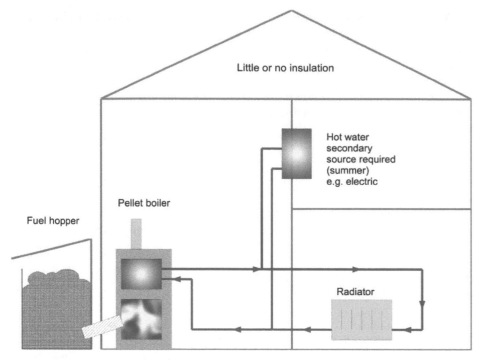

Figure 4.15 *Replacing total heating system with wood pellet boiler in typical older house*

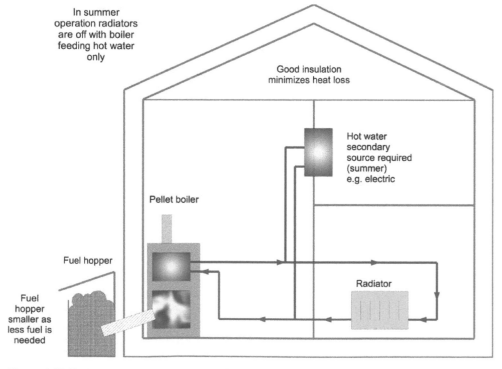

Figure 4.16 *Replacing total heating system with wood pellet boiler in low-energy house*

- fuel hopper;
- burner and combustion chamber;
- fan-assisted warm air circulation;
- back boiler for heat exchange to domestic hot water (via cylinder) and/or radiator or underfloor heating.

Stove with back boiler and solar thermal input
An increasingly popular and very green option, this could be ideal for a small home or for a holiday cottage occupied mainly in summer when the majority of hot water demand will be satisfied by the solar thermal input on its own. This improves overall annual boiler efficiency by reducing the amount of cycling required during the summer months. The main features of this system (see Figure 4.17) inevitably include:

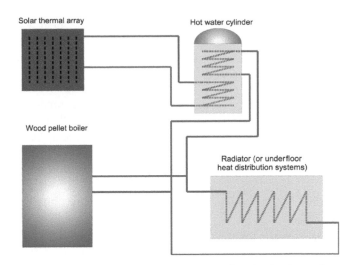

Figure 4.17 *Boiler (or stove with back boiler) and solar thermal input*

- fuel hopper;
- burner and combustion chamber;
- fan-assisted warm air circulation;
- back boiler for heat exchange to domestic hot water via cylinder and/or radiator or underfloor heat distribution system;
- array of solar thermal panels.

Boilers

Pellet boilers now claim up to 90 per cent efficiency and can provide central heating or process heat for most requirements from the home to the factory. Furthermore, they can either replace existing boilers (oil, gas or coal) or be installed alongside them to operate in conjunction with them even without other major alterations to the heating system. Most pellet boilers use a two- or three-stage combustion process and ceramic thermal lining. Efficiencies of pellet boilers are usually aided via a buffer tank.

There are a number of ways to classify boiler types, but in terms of their construction there are two principal methods: cast-iron sectional and welded sheet metal.

Cast-iron sectional pellet boilers offer the advantage of robustness and ease when transporting and installing, since they can generally be put together on site from several prefabricated component parts. This is particularly useful if the boiler is to be installed in an internal room or basement.

Usually less expensive, sheet metal boilers are welded together in the factory, so they are always delivered as one integral unit, which means that installing it is a common problem.

Wood pellet boilers operate on negative pressure, drawing air into the burner and through the combustion chamber. This is achieved either via a motorized fan or by natural draught and ventilation flows. Flue gas fans can be positioned between the boiler and the flue in pellet boilers that have too much gas resistance in the heat exchange zone. These fans can have adjustable controls. In boilers with less gas resistance, natural ventilation can usually be achieved by having a tall enough chimney.

Domestic boiler systems

The heat outputs of compact boilers available on the market today range from 750W up to around 50kW. Common configurations include connection to new or existing sealed boilers. Unsealed systems can be used, although most pellet boiler manufacturers produce boilers with sealed systems in mind. Solar thermal works well in conjunction with a pellet boiler system and many buffer tanks are designed to allow for this eventuality, as is much pellet boiler control software.

Even though these units are known as compact boilers, they take up more space than an oil or gas boiler of comparable output. The ÖkoFen PELLEMATIC range starts at 8kW and goes right up to 56kW, yet the difference in floor space footprint between these extremes is not as much as one might expect, with the 8kW unit at 691mm (27 inches) by 1013mm (39 inches) and the 56kW unit at 1015mm (40 inches) × 1297mm (51 inches). The exhaust tube diameter for flue connection is also only marginally different at 130mm (5 inches) for the 8kW and 180mm (7 inches) for the 56kW.

Commercial boiler systems

Available wood pellet boilers for commercial applications range from around 50kW right up to 1MW. Even at the lower end of this range, most pellet boilers are installed with factory-programmed controls that cater for a host of possible configurations. KWB's boilers, manufactured in Austria, are designed to manage up to 34 heating circuits, along with 17 domestic hot water outlets and 17 buffer tanks.

Pellets are used in much larger heating and CHP plant (up to 250MW), particularly in Scandinavia, most commonly mixed with wood chip and/or coal

Table 4.1 ÖkoFen boiler data

Boiler type		PE08	PE12	PE15	PE20	PE25	PE32	PE36	PE48	PE56
Boiler capacity (nominal)	kW	8	12	15	20	25	32	36	48	56
Breadth (total)	B mm	1013	1130	1130	1130	1186	1186	1297	1297	1297
Breadth (boiler)	C mm	645	700	700	700	756	756	862	862	862
Height (boiler)	H mm	1066	1090	1090	1090	1290	1290	1553	1553	1553
Height (vacuum system)	D mm	–	1520	1520	1520	1710	1710	1855	1855	1855
Depth (boiler)	T mm	691	814	814	814	870	870	1015	1015	1015
Depth (burner cladding)	V mm	430	508	508	508	508	508	508	508	508
Flow and return (diameter)	Inch	1	1	1	1	1¼	1¼	2	2	2
Flow and return (height of connection)	A mm	896	905	905	905	1110	1110	1320	1320	1320
Exhaust tube (diameter)	R mm	130	130	130	130	150	150	180	180	180
Exhaust tube (height of connection)	E mm	664	645	645	645	844	844	1040	1040	1040

Source: ÖkoFen

for combustion. At this commercial scale, several boiler types are available on the market: grate boilers, compact boilers and boilers with detachable stokers.

Back-up boiler in commercial property
In commercial systems, which often depend on heat for processes as well as for space and water, back-up boilers are sometimes used alongside wood pellet boilers. Both are usually plumbed into the same heat distribution system. Their use doubles up as a back-up in case of main pellet boiler breakdown, or during routine maintenance, and they can also be switched on to meet peak loads in the middle of winter. In some circumstances, the back-up boiler may also be sized to operate alone at the margins of the season, with a larger main pellet boiler kicking in as the temperatures drop below the capability of the secondary boiler. Back-ups are generally oil- or gas-fired boilers, although pellet boilers can also be effective. Using pellet boilers as back-up can also mean savings because fuel for both can be received and stored in the same way and only one chimney is needed; but they will still need independent fuel delivery and feed mechanisms from the pellet store to the boilers.

If a pre-existing oil or gas boiler is efficient and appropriate, it can remain in place to serve as a back-up. Specifying a new back-up boiler at the system design stage, however, is always more advisable, permitting a tighter design compatibility and almost certainly greater economies in the longer term. A rule of thumb when using a wood pellet boiler to supplement another pellet boiler is that one should cover the base load (about 70 per cent of the heat load) while the second should offer additional input (30 per cent) to cover peak and low loads.

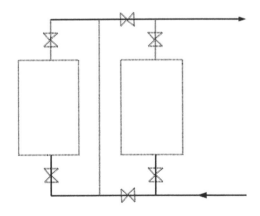

Connecting boilers in parallel or series
Where there is more than one boiler operating in a property they can often be connected to one another either in series or in parallel. Series connections, as in Figure 4.18, can suit situations where the smaller boiler has a relatively low heat output, significantly lower than the base load, but uses a fuel which is more expensive than wood pellet (for example, an electric, coal, LPG or oil boiler). In these systems, the water flows through first one boiler then the next, unless controlled via the inclusion of valves. Where boilers are connected in parallel, as in Figure 4.19, they both receive the flow of water at all times, even when one of the boilers may not be switched on.

Figure 4.18 *Connecting boilers in series*

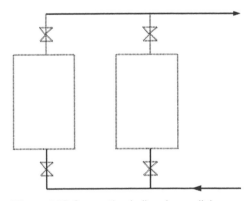

Retrofitting a pellet burner into an existing boiler
Retrofitting a pellet burner into an existing oil or coal boiler has proved an attractive proposition, particularly in Scandinavia during the early days of wood pellet technology development. New burners must be sized and positioned to fit precisely the old boiler's heat exchange unit and design. This is most likely to be an option where there's an existing quality boiler in good operational condition.

Figure 4.19 Connecting boilers in parallel

The disadvantages are guaranteeing the reliability of a cobbled-together boiler and the availability of trained technicians for maintenance and trouble-shooting. Although this retrofitted burner can make the overall system significantly less expensive, it is important that it can be easily dismantled for the frequent cleaning that will probably be required. Requiring similar levels of skill to designing a wood pellet system from scratch rather than purchasing a system complete on manufacture, retrofitting burners with boilers that have a record of compatibility is always wise.

Accumulation tanks

Accumulation tanks are used in larger or more complex wood pellet boiler systems to help manage variable heat demands. They can add preheated water into the system for rapid response to increased loads and also provide hot water during periods of low heat demand, saving the boiler from unnecessary cycling.

If a boiler system is supplemented by solar thermal input, then an accumulation tank or some hot water buffer store is a very important element. The size of the buffer tank is a factor of the efficiency and surface area of the solar panels. Solar water heating is highly compatible with wood pellet heating systems. Solar thermal systems receive most of their solar heat during the summer, often avoiding the need for the wood pellet boiler to switch on simply to heat domestic hot water for a few months of the year. There is still sunshine sometimes even in the depths of winter, and this solar thermal input can also be used through the accumulation tank to offset some of the space or water heating needs at this time, again saving the boiler from additional exertion.

Pellet storage and automatic fuel feeds

Alongside the environmentally friendly nature of pellet technology, convenience for the user-operator and full automation are arguably the most desirable features of wood pellet systems. Pellet stoves are designed to burn a fuel that moves in some ways similar to a liquid product, flowing down ducts and moving without blockages or manual intervention into the combustion zone. The key to this convenience is clearly the wood pellet itself. Pellet size, shape, consistency and content are all vital for enabling fuel flow. And if they are poorly stored, even the best-quality pellets can turn to mulch or dust.

Once a reliable supply of quality wood pellets that can be delivered to the premises to be heated has been secured, the next thing that needs to be ready is a clearly thought-out and well-prepared pellet storage area or tank. Even stoves that are only occasionally used will need somewhere to store a 10kg (22 pound) bag or two of pellets, ideally off the ground, away from walls and any dampness. But for pellet boiler systems, the fuel storage options are probably the first concern, even before selecting boiler and system types. Pellet fuel storage does take up significant space.

Common pellet store locations and types

There are a number of key issues to weigh up when thinking about pellet store location and design. Perhaps the most important consideration, though, is how the pellets are going to be delivered. In most cases where there's a pellet boiler, the choice will be bulk delivery by blower lorry. But there are several other general issues:

- **Air pressure during blowing** – Any pellet store that is filled by blowing pellets in must always have at least one air vent of sufficient surface area to ensure that the doors and windows are not blown out under the air pressure, which is inevitable as the pellets are blown into a confined space.
- **Air flows** – A good flow of air is recommended in all cases just to keep the pellets aired and dry.
- **Strength of footings** – If a shed is large enough to fit a pellet storage tank (see below for illustrations and more details), then it is important to ensure that the footings for the tank's feet are designed for the weight and forces they will experience when the tank is full.
- **Fire risks** – There is always a fire risk with any fuel store, so care must be taken to minimize electrical or burn-back risks (see Chapter 5); fire extinguishers should be kept visible and within easy reach in case of an emergency.

Dry external shed

Garden or other sheds that are dry are often used to store plastic 10kg (22 pound) or larger bags of pellets. If the shed has good foundations and a floor that can take the weight of a larger pellet store, then it is sometimes possible to use them for bulk pellet storage. Because of the fluid nature of wood pellets, however, it's not possible to blow a ton or two of pellets into your average shed, even if there's enough space. The shed will collapse outwards with the sheer weight and outward force of the pellets against the walls.

Dry internal cupboard

This type of storage is only suitable for the occasional stove user or for stoves that are not needed to provide a property's base heating loads, but to supplement a larger heating system. A few bags can be stashed out of the way in a dry cupboard. It's always sensible to try to keep the bags off the ground and away from the walls, just in case of flooding, condensation problems or rising damp. When the user needs to refill the stove, they simply carry a bag and pour it into the in-built hopper.

Outbuilding

If the property to be heated with pellets has an appropriate outbuilding already there, then this is often a cost-saving option. The same precautions should be taken as with an external shed. They must be dry, or it must be possible to store the pellets away from damp and water, with a good air flow. Ideally, the outbuilding will be attached to the room where the wood pellet boiler is going to go, otherwise automatic delivery of the pellets may be impossible. If there is a gap between the outbuilding and the property to be heated, then the pellet boiler itself should be sited in the same outbuilding and super-insulated pipes can take the hot water from the boiler into the main building, acting as a mini heat main.

Boiler room

Boiler room and fuel storage spaces should always be discussed with your boiler supplier in advance of design. If there's an existing boiler room with sufficient space and access for bulk delivery by blower, this will be the first place to assess for possible pellet storage. The same constraints as above apply regarding air flow, dryness and strength of floors and foundations. Furthermore, the angles and distances needed for the fuel delivery and feed mechanisms need to be taken into account. It's a good idea to draw a diagram of the new wood pellet boiler's footprint and those of auxiliary systems such as the store tank and fuel delivery to make sure there will still be enough space to work around in the boiler room. Wood pellets are a fire risk in boiler rooms and should never be stored loose outside a container in this situation. Chapter 5 provides more information on this.

Purpose-built store

If there's no suitable shed, boiler room or existing outbuilding available, then a purpose-built pellet store is always an option. Ground space and delivery access are obviously necessary and all of the same precautions as above apply. But it is particularly important to position the store directly on the other side of the wall from the boiler, otherwise, an automatic auger delivery mechanism could well be very difficult to fit because it will have to go through one of the property's outer walls and may be angled both horizontally as well as vertically. There is more detail on the pellet delivery mechanism options below.

Prefabricated boiler/store rooms

Prefabricated boiler or store rooms (for example, the ÖkoFen Energy Box; see Figure 4.20) are an increasingly common energy solution, particularly for larger applications such as schools and hospitals, but are also available at the domestic level. There will be more on this in later sections of the book, but, in brief, these are sheds built from wood narrow enough to be delivered by lorry and arrive with the boiler and pellet storage tank ready for plugging into the property. They can be dropped off by lorry next to the building, connected to the electricity and plumbing, then simply switched on.

Tanks and other containers

Most pellet boiler systems will store pellets loose in some kind of tank or container rather than on the floor, wherever this may be. It is possible to use an underground silo for pellet storage, which is particularly appropriate if the boiler itself is in a large basement room. Whatever the container, it will ideally be designed to hold your entire winter store (see Chapter 5 for sizing stores). This can keep the fuel cost down since only one delivery is necessary each year, and there are sometimes price advantages to buying at specific times of year (for example, the summer months).

Some manufacturers make and install their own flexible tank system for wood pellet fuel storage up from 450kg (992 pounds) to 7 tons, such as

Figure 4.20 *ÖkoFen Energy Box being installed at a UK housing association to provide district heating*

Source: Organic Energy, 2009

ÖkoFen's FlexiTank (see Figure 4.21). These are designed for use with either vacuum suction or auger screw pellet delivery mechanisms. Other similar tanks or silos are available on the market. The materials used are generally tried and tested. It is possible to build one yourself, but it's very important that the structure can take the weight, is compatible with the pellet delivery mechanism and can take the pressures of blowing pellets into it and that the fabric used is strong and dense enough to contain the pellets and pellet dust.

Automatic pellet fuel-feed conveyance systems

There are two main types of such conveyance systems in use: one which operates by vacuum suction, the other by moving the pellet fuel via an auger screw mechanism. Both generally deliver the pellets straight into the boiler, either via an internal intermediate boiler hopper, or direct from the auger, often using

Figure 4.21 FlexiTank pellet storage system

Source: ÖkoFen

another smaller auger (the stoker auger) to deliver the fuel right into the burner. There will be more on this in Chapter 5 when we look at designing entire wood pellet boiler systems, from pellet reception to boiler and heating circuit configuration.

Vacuum suction conveyance

The maximum distance recommended for pumping pellets is 20m (65 feet). Shorter pumping distances are always preferred. The pellets are simply sucked from the tank or silo to the boiler as directed by the boiler controls. It's important not to have too many bends and any acute angles in the piping, so as to avoid pellet blockages.

The vacuum suction conveyance system is particularly appropriate for situations where the storage silo is in an adjacent or separate outside building (see Figures 4.22 and 4.23). It is easier to retrofit a vacuum pipe than it is an auger screw mechanism.

Auger screw conveyance

Auger screw mechanisms are the most common and generally most economical conveyance system to run. Their main disadvantage is that because of the length of the auger screw mechanism, the silo must be adjacent to the boiler.

While it is possible in some situations to send the auger through a hole in a wall (see Figures 4.24 and 4.25), this is not always feasible, simple to achieve or appropriate.

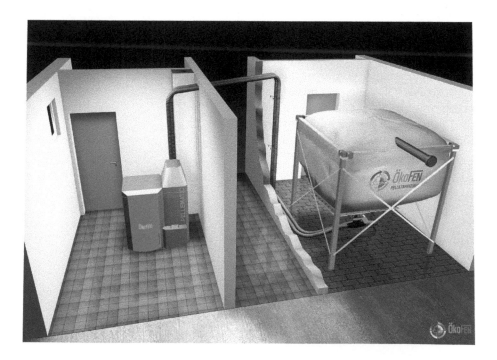

Figure 4.22 FlexiTank with vacuum suction conveyance

Source: ÖkoFen

Wood pellet delivery and distribution

Wood pellet delivery

Even 10–25kg (22–55 pound) bags of pellets can be stored in stacks on pallets. In Sweden, bulk bags containing between 500kg (1102 pounds) and 800kg (1763 pounds) can be delivered to households. Most boiler users, however, will want the bulk lorry delivery that can blow pellets from mobile storage tanks directly into their store, making the delivery process fully automated as well. The lorry will need to be able to park near enough (20m (65 feet) maximum, preferably 15m (49 feet) or closer) to the storage area. The correct design for the store tank to blower connectors is discussed in Chapter 5. Where there's an underground silo, a tipper lorry can deliver pellets into this, but the silo lid opening must be large enough to receive them without spillage.

Wood pellet distribution

Traditionally, wood pellets are loaded into a bulk lorry at the pelletization factory and driven to the consumer's premises. Many of these trucks have segmented tanker capacity permitting three or four separate tanks of pellets in each and

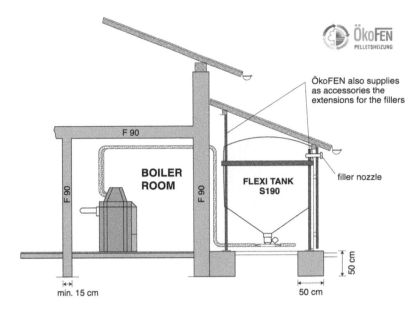

Figure 4.23 *Cross-section of FlexiTank with vacuum suction installation*

Source: ÖkoFen

Figure 4.24 *FlexiTank with auger screw conveyance*

Source: ÖkoFen

enabling more than one delivery to take place. Having separate tanks also means that, when there's an empty tank, the lorry can turn the blower into a vacuum suction device, removing the dust in your store before filling with pellets from another separate full tank. Most blower trucks have a maximum range of 30m (98 feet) for blowing, although bends and constraints restrict this in reality; however, their ability to suck back pellets from a store is restricted to around 20m (65 feet) (less with constraints).

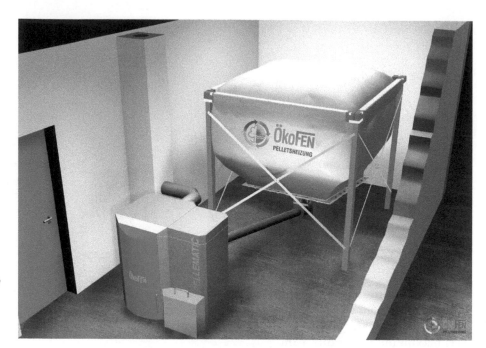

Figure 4.25 *Cross-section of FlexiTank with auger screw device for pellet conveyance*

Source: ÖkoFen

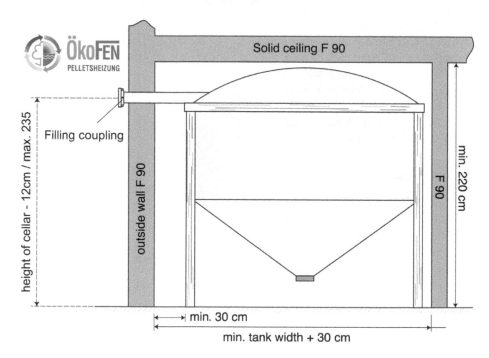

Figure 4.26 *Cross-section of pellet store tank with blow-filling coupling to the outside wall*

Source: ÖkoFen

In emerging markets such as the UK, bulk pellet delivery by lorry has sometimes been taken up by the animal feed transport industry. But in Austria and Germany, large pellet manufacturing companies have their own fleet of trucks (see Figure 4.27). Many of them find it economic to deliver to customers within a 500km (310 miles) radius.

Figure 4.27 *Pellet delivery vehicle*

Source: Balcas

System efficiency issues and heat distribution

System efficiency issues

With the introduction of condensing wood pellet boilers, the potential efficiencies can now be pushed well over 93 per cent, but most pellet boilers hover between 85 per cent and 90 per cent. There are a number of pellet boiler system elements that affect overall system efficiency. Each of these is covered in details in Chapters 5 and 6, when we look at design and installation. For the moment, however, we offer a quick overview of the main issues.

Electrical inputs

Heat distribution pumps aside, both stoves and boilers require an electrical connection, usually for ignition, motorized fuel feeds, exhaust flue fan, warm air distribution fan (in stoves) and sometimes a combustion fan.

Pellet quality

As the basic fuel, if this is of poor quality in terms of consistency, dryness or constituent materials it will make the system run less efficiently, possibly snagging up the feed auger or being lost as dust, and could cause soot accumulation and unwanted emissions.

Combustion design

Design of the burner plate, feed mechanism, combustion chamber and heat exchange units will all impinge on efficiency. Most manufacturers have technical specifications and trial data available for their boilers and these should be carefully compared at boiler selection stage.

Intelligent modulating controls

This is the clever part of any boiler and is of paramount importance for determining overall system efficiency. If the boiler can modulate smoothly to suit user hot water and heat demand, this will save on fuel and maximize the heat output from that used.

Heating system type

The choice of open-vented or sealed system also brings up efficiency issues. Sealed systems are generally more efficient with smaller pipe runs and no tanks open to air within the system.

Optimum boiler sizing

If you select a boiler that is too small it will have to run flat-out all the time, which shortens the life of the apparatus and can be costly on fuel. It may not even get the room temperatures up to the desired levels on cold days. A boiler that's oversized will end up operating at the lower end of its capacity, probably having to modulate and cycle excessively, again generating unnecessary inefficiencies and fuel costs.

Multi-boiler configuration

It may be more efficient in larger premises to fit two boilers in series or parallel. With one taking the base load and the other meeting peak and low demand, if the ratios are right for the property and the way it is used, this can be very cost-effective in terms of fuel use.

Use of a buffer tank

To help reduce modulation and cycling in the boiler and to provide instant hot water that can be mixed to the right temperatures for heating or hot water demands, a buffer or accumulation tank is a sensible option for most pellet boiler systems.

Solar thermal input

Similarly, if the property has a good solar aspect, it is worth considering investment in an additional thermal input in the form of an array of solar thermal panels that can be plumbed directly into the buffer tank, so it can share its heat with the rest of the system.

Flue design

As with all wood-fired stoves and boilers, the height and draught action, or pull, of the flue and chimney system will affect efficiency and unless the boiler controls can adjust for this, the issue may need to be managed at the design stage. If the pull is too great, reducing chimney height may be one way of stopping too much fuel being burned. If it's too weak a pull, then it's usually possible to improve this by raising the level of the chimney top.

Heat distribution

The two main wet heat distribution systems used by wood pellet boilers are radiators and underfloor systems. A relatively new product on the market, ThermaSkirt, offers a third option.

Radiators

Metal radiators of one sort or another are the most common device for wet heat distribution. Hot water from the boiler is pumped through pipes into radiators that are these days usually made from sheet steel. Inside the radiator, the hot water circulates and gives out much of its heat to its metal fins or plates (usually steel), which are designed to increase the surface area of the apparatus and transfer heat to the surrounding air. Wet radiator systems generally operate at temperatures between 60°C (140°F) and 80°C (175°F); if they are to operate at temperatures of 40°C (105°F) to 60°C (140°F) they would need to increase the surface areas of the radiators by some 30–40 per cent to have the same overall heat output. So, generally speaking, big radiators require lower circulation temperatures.

Underfloor

Underfloor coils are an increasingly common option for wet heat distribution. Best suited to ground floor areas, and ideal in a new-build scenario, a circuit of coiled pipes can be laid under a floor. Often laid on top of sand under a concrete or tiled floor, such a system allows heat to rise from the fabric of the floor itself into the room. It can be very comfortable to live with, although some people complain that it makes the air too dry. The hot water in underfloor systems usually runs at much lower temperatures than in radiators. They can work well with either a pellet stove with back boiler or as part or all of any pellet boiler's heat delivery system. Another advantage of underfloor heating is that it takes up absolutely no wall space.

ThermaSkirt

Designed and manufactured by The DiscreteHeat Company, ThermaSkirt is a product that replaces radiators and skirting boards in one. Like underfloor heating, it heats a room from all sides and at low temperatures. It is wholly compatible with wood pellet boiler systems and claims 13 per cent more efficiency than radiators. Also unlike radiators, ThermaSkirt allows for ample furniture space along all walls. It uses a unique aluminium alloy material that is said to be five times more effective than steel at passing heat and can run efficiently at temperatures between 35°C (95°F) and 75°C (165°F).

Wood pellet CHP and use in CHP co-firing

At the larger scale, wood pellet fuel is used for CHP generation, getting the most out of the fuel by generating electrical power as well as useful heat. Co-firing biomass is something that the Obama government is looking at as a way to reduce emissions from coal-fired power stations, given that wood pellets can be directly co-milled and burned with coal in existing mills and burners. In the USA, co-firing with biomass has the technical and economic potential to replace at least 8GW of coal-based generating capacity by 2010 and as much as 26GW by 2020 (Morand, 2009b). Much of this could use wood pellets.

Wood pellet CHP

CHP generation from wood pellets has been common in Sweden and Finland for many years. It has also emerged in the USA, Canada, Austria and Germany. One company in the UK has been set up to generate power from burning wood pellets that it plans to produce on the same site from local forestry thinnings, which are otherwise difficult to market. Creating a new market for thinnings can therefore help fund and encourage good management of forests. The thinnings will be dried by using the heat element of the CHP unit. Although possibly a good idea, it could be seen to have two downfalls environmentally. First, the heat will not all be required for the drying process and could arguably be used in a more valuable way for other processes or properties. Secondly, it is arguably better and certainly less energy-intensive to use a sawmill co-product rather than tree thinnings, since the co-product is already part-dried and part-pulverized.

By the mid 1980s, Sweden was already producing over 50,000 tons of wood pellets a year. In the early 1990s, the country introduced a new tax on fossil fuel carbon emissions, which, combined with the development of pellet boilers of up to 100MW capacity, really got the pellet market going. At Hässelby, in the northeastern part of Stockholm, on the shore of Lake Mälaren, a massive heat and power plant is Sweden's largest user of wood pellets, consuming 250,000–300,000 tons a year. Pellets are delivered to Hässelby by boat, and the plant was one of the earliest to convert from fossil fuel to biomass. Hässelby was a starting point for large investments in pellet manufacturing across Sweden.

Pellets and co-firing

Co-firing is the combination of fuels, usually for the generation of heat and power. Pellets are mainly used for co-firing across Europe in Nordic countries, The Netherlands, Belgium and the UK. In the USA, during June 2009, the energy firm Green Energy Resources, one of the continent's leading companies in the provision of low-cost, high-quality, high-energy wood fibre fuels, received two separate orders totalling 20,000 tons of wood pellets (estimated to be worth US$28 million) to be co-fired with coal up to the end of 2010. Much larger amounts are already being co-fired in Europe, and the Drax plant in the UK anticipates consuming around 1 million tons of pellets a year in the near future. The EU currently supplies over 4 per cent of its total electricity from wood waste. This compares with around 2 per cent in the USA. Until now, European policy instruments and the Swedish abundance of pellets have driven world demand for large scale use in co-firing. This number is expected to double in 2010, with present consumption at over 6 million tons a year (Morand, 2009a).

Solar-assisted systems

Having a wood pellet boiler or stove with back boiler running in the summer just to provide domestic hot water is a waste of fuel. Much greener and more economical in the long run is to match a wood pellet system to an appropriate solar thermal array. This works out cheaper and is easiest to install at the same time as the boiler and central heating system with which it is integrated through the hot water cylinder or buffer tank. The two systems cannot be combined effectively without a well-insulated cylinder.

Solar thermal panels (also known as collectors or absorbers) heat a liquid circulating in an absorber. As it gets hotter, the fluid is pumped through pipes into a coil inside the hot water cylinder or accumulation tank.

Special solar cylinders are available that maximize efficiency by having specially designed and placed internal coils, one for the solar input and at least another for the boiler (see Figure 4.28). Sophisticated pellet boilers can also manage the solar hot water through the buffer tank and pumping stations to meet space heating demands. The size of the buffer tank needs careful consideration in relation to the property's hot water and heating demands, their variability and the solar radiation available at the specific location. These issues are considered in Chapter 5.

Many wood pellet boiler manufacturers have adopted or manufacture solar systems whose design is totally compatible with their boilers. ÖkoFen, for instance, manufactures the PELLESOL solar array (see Figure 4.29) combining high quality design certification from Solar Keymark.

Solar arrays can be fitted on top of a roof or integrated into it and consequently lying flatter (see Figure 4.30). Either way, they are highly efficient in hot water production. According to the manufacturers, just 1.5m² (16.15ft²) of collector area can generate around 70 per cent of the hot water required per person per year.

Some pellet boiler manufacturers also produce their own buffer tank specifically for use with wood pellet boilers that have some solar thermal input. Programming for a solar array comes as standard in each boiler's controller set-up.

Many accumulator tanks have integrated heating controls (see Figure 4.31). These provide instantaneous hot water from the mains and are generally supplied ready for installation with all fittings and electrics. They are often designed to work with integrated pump stations and more than one heating circuit. Optional extras usually include a solar water heating pump and an accumulator loading pump as well as an integrated solar heat exchanger.

Solar heating of domestic water became a growth market following the first oil crisis in 1973, after which environmental awareness and energy insecurity both grew. According to the International Energy Agency (IEA), by 2004 the global installed base of solar thermal systems had an output of roughly 70GW (International Energy Agency, 2004). But there is clearly much more potential. Today's systems are powerful with highly efficient collectors that can respond quickly to changes in weather thanks to sophisticated control technology. When the sun simply is not available, then the wood pellet boiler will switch on automatically to satisfy any demand. In a country such as Austria, solar collectors

Figure 4.28 *Cutaway of solar cylinder*

Source: Solarcyl

are frequently used for space heating in combination with wood pellets. Large solar thermal arrays, sunny winters and sufficient capital resources are required for solar space heating.

Figure 4.29 *Cross-section of typical solar array*

Source: Organic Energy

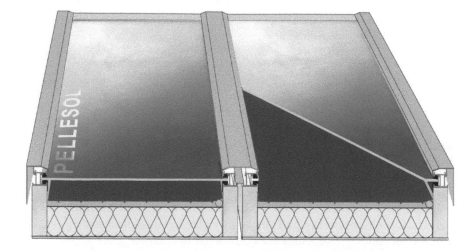

Figure 4.30 *A roof-integrated solar thermal installation*

Figure 4.31 *PELLAQUA accumulator tank*

Source: Organic Energy

References

International Energy Agency (2004) 'Worldwide capacity of solar thermal energy greatly underestimated', www.iea-shc.org/welcome/Press%20Release%20-%2070GW%20solar%20thermal%20capacity.pdf (accessed 5 January 2010)

Morand, C. (2009a) 'Investing in Wood Pellets, Part I', altenergystocks.com, 19 March, www.altenergystocks.com/archives/2009/03/investing_in_wood_pellets_part_i_1.html http://bioenergy.checkbiotech.org/news/investing_wood_pellets_part_i (accessed 20 November 2009)

Morand, C. (2009b) 'Investing in Wood Pellets, Part II – A Stock', altenergystocks.com, 2 April, www.altenergystocks.com/archives/2009/04/investing_in_wood_pellets_part_ii_a_stock.html (accessed 20 November 2009)

5

Designing and Sizing Wood Pellet Heating Systems

Design principles and general methodology

Principles

As with planning any project, preparing to install a pellet boiler – whether it's to provide heat for anything from a family home to a factory or industrial process – requires clarification of the main objectives. Usually, this is a combination of hot water provision and thermal comfort in a building within the constraints of a given budget.

The next task – working out how much heat is required to do this – is critical. Once the heat load has been defined and refined, then the boiler (or boilers) can be sized properly.

In simple terms, the process involves the following steps:

A   List spaces for heating.
B   Work out floor area of spaces to be heated.
C   Establish ideal temperature for each space to be heated.
D   Use figures B and C (plus any information on building heat loss or air changes) to work out heat requirement for each space. The results can be used later to aid in specifying radiators.
E   Add these to give total space heating load (if they are particularly tall rooms – e.g. over 3m (10 feet) – then an adjustment upwards must be made).
F   Add in other load requirements for the property (e.g. domestic hot water).
G   E + F = BASIC HEATING LOAD.
H   Take into account other considerations (e.g. other significant thermal inputs from a solar thermal water collector, expected simultaneity in heat demands or lots of lighting emitting radiant heat, as well as dynamic effects).
I   G – H = IDEAL BOILER CAPACITY.

For a property owner and potential wood pellet boiler user, going through these steps will provide a working notion of the required capacity. But it is always very important to get your boiler manufacturer, authorized dealer and/or qualified installer to go through the process according to their own techniques and methodology. After all, they are the ones offering a warranty and if they are any good they will already have lots of experience at sizing pellet systems.

Some of the more advanced boilers are designed to anticipate the heat loss and ensure smooth and efficient operation through a sophisticated heating control strategy combined with variable temperature heating circuits, often in combination with hot water storage systems. The efficiencies gained mean that they often specify much smaller boiler capacity than property owners are expecting. Too large a boiler will result in poor combustion quality due to frequent shut-down and start-up cycles as well as partial loads. It also incurs unnecessarily high initial investment costs and can sometimes be responsible for woodsmoke smells in the local neighbourhood. The smallest boiler possible to deliver the heat needed is generally considered the right choice (if necessary, with a hot water storage or buffer tank); it may also have less of a need for flue gas cleaning.

Methodology

The steps in designing a wood pellet boiler system can be divided into two main phases:

- **Phase 1** – Preparation: this stage looks at financial issues, health and safety matters, heat loading, space constraints, boiler sizing and options etc.
- **Phase 2** – Specification: this stage covers boiler and auxiliary system component selection, heating circuit design, level and sophistication of controls etc.

Installing a wood pellet boiler is a project in its own right and the design and installation should be approached as one would manage any other complex project, with planned timings for the main tasks. This is particularly important at the equipment delivery and installation stage (see section on project development and community projects below). There are various software tools for project management available, but it can just take a simple spreadsheet or detailed list with timings. However, it's important to remember to check the list regularly throughout the design and installation process.

Phase 1 – preparation

Establish objectives

Usually, the objectives are to attain certain thermal comfort levels and a specific amount of hot water provision, all when needed. The level of automation and carbon-neutrality are other common criteria for selecting a wood pellet heating system. If there are specific stylistic requirements, these too should be incorporated.

Heat loss calculations

For a new-build heating system design, it's possible to apply general rules based on material specifications for the building. When it comes to older buildings, dimensioning your boiler according to hard historic fuel consumption data is always a pretty reliable method (for example, by producing a heat load duration curve). To complement this, though, given the capital costs involved, it's a good idea to have a professional survey carried out that includes heat loss calculations based on a range of factors (see list below).

It is always advisable to involve a professional energy auditor for a room-by-room assessment of the premises and an analysis of past fuel and utility bills. A thorough audit could include a blower door test (a powerful fan that pulls air out of a building to create exaggerated but detectable drafts). Some may also involve an infrared thermographic scan or even a PerFluorocarbon tracer (PFT) gas air infiltration measurement to get to grips with air leakage and infiltration (which is excellent for longer-term analysis).

Given local climatic conditions, the boiler capacity is mainly determined by the heat output demanded by the property. This can be calculated on a room-by-room basis, selecting radiators of specific outputs to satisfy heating requirements in the different spaces. Radiator outputs are standardized by kW or British thermal unit (BTU) rating. For instance, if you have three rooms each with a 4kW radiator and two with 3kW radiators, then the peak space heating requirement can be stated as 18kW. Domestic and any other hot water requirements also have to be calculated and added in to get to the total heat output demanded by the property. If water heating requirements are estimated at 2kW peak, then the total heat demand should be 20kW. This is a good indication of the size of boiler needed for the task in hand.

The key factors influencing heat loss are as follows:

- local climate (temperature averages and extremes plus winter length);
- orientation of premises (particularly at building exposure);
- size, type and shape of premises (one or more storeys, detached or terraced, surface areas facing exposure issues, etc.);
- insulation levels (roof, walls, floor);
- window area, locations and type;
- rates of air loss (or infiltration);
- building occupier behaviour and level (number of occupants, lifestyle pattern relating to heat and hot water use; these are things that can obviously vary greatly, for example, between a care home and a townhouse occupied by a single person who works and spends leisure time away from home);
- occupant comfort preferences (room temperatures and use patterns);
- simultaneity in heat demands;
- dynamic effects (sudden heat demands, such as when a person enters a room);
- other heat inputs to premises (significant solar gain from a large south-facing window or unheated conservatory, solar thermal collectors plus types and efficiencies of lights and major home appliances, since all of these give off varying amounts of heat).

Table 5.1 *Suggested target temperatures by room*

Room	Temperature (°C)
Main living and eating areas	20–21
Bedrooms	15
Kitchen	16
Bathroom	22–23

In the USA, a commonly used reference work for domestic installations is the *Manual J Residential Load Calculation* published by the Air Conditioning Contractors of America (ACCA). For the UK, *Domestic Heating – Design Guide* is a very useful booklet, published by CIBSE. Wherever a unit is located, property owners should always ensure that contractors and/or pellet boiler suppliers use a correct sizing calculation before signing a contract. It's a good idea to get more than one potential supplier to go through this process for you. The comparison of different suppliers' sizing results could provide important information. When the contractors are finished, always get a copy of their calculations. If you want to be better prepared before calling a supplier or contractor, then boiler sizing services can often be accessed for a fee through most utilities and good heating contractors.

It's always easier to accurately estimate boiler requirement for new-build properties where the building area and insulation levels are already known. The other factors will have to be assumed. If a district heating network (DHN) is involved, then heat losses should be calculated for this according to the heat main's predicted heat loss per metre (these figures should be available from the heat main supplier or contractor).

Predicted use patterns (user behaviour)

Essentially, these patterns will determine the room temperatures for premises (see Table 5.1). With sophisticated controls and zoning they can vary from space to space and time to time. An average use pattern can usually be calculated by making intelligent assumptions about predicted use.

Energy efficiency improvements

DIY energy audits can usually pick up many problems in any property. Keep a checklist and inspect room by room as well as outside to help get an accurate picture and also prioritize any energy efficiency improvements that may be needed. Start with draughts and air leakage (these can be responsible for up to 30 per cent of heat loss) along floors, edges of rooms, corners and ceiling, as well as window frames, external doors, electrical outlets, hatches, letter boxes and so on. For those having difficulty locating leaks, the US DOE recommends conducting a basic building pressurization test simply by closing all exterior doors, windows and fireplace flues, turning off all combustion appliances, then turning on all exhaust fans (generally located in the kitchen and bathrooms) or using a large window fan to suck the air out of the rooms. This test increases infiltration through cracks and leaks. Incense sticks or small smoke detectors can be used to find any leaks as can your hand (especially if dampened).

Heat loss through ceiling, walls and floor can be addressed. Lofts and roofs lose up to 25 per cent of a building's heat if uninsulated. In the UK, the recommended depth for mineral wool insulation is 270mm (10 inches), although other materials require different depths. The insulation levels of exterior walls are best judged by their material constituents and construction (see Box 5.1).

Box 5.1 U-values and R-values

U-values, commonly used in the UK, represent the overall heat transfer coefficient, which means that they describe the rate of heat transfer through a building material over a given area. R-values are the reciprocal of U-values, since they measure thermal resistance in buildings and building materials. They are usually stated as the ratio of temperature difference across an insulator and heat flow per unit area. Higher R-values and lower U-values both represent more effective insulation. Recommended U-values (or factors) describe a recommended maximum, while recommended R-values specify a minimum.

U is the inverse of R with units of watts, metres and Kelvin (or degrees Celsius) rather than the US R units, which are based on degrees Fahrenheit and BTUs.

In the USA, the Federal Trade Commission (FTC) governs claims about R-values to protect consumers. Elsewhere, R-values are given in SI units: $m^2 \cdot K/W$ (or $m^2 \cdot °C/W$):

- U-value (UK) = thermal conductivity properties (W/m^2K)
- R-value (USA) = thermal resistivity ($ft^2 \cdot °F/BTU$)
- SI = thermal resistivity ($m^2 \cdot °C/W$).

SI and US R-values can be confused when cited without specifying the units (e.g. R-3.5).

Other renewable energy inputs

The most likely option here is attaching a solar thermal collector to the boiler system, since they are so seasonally compatible (for detailed information see Earthscan publication *Solar Domestic Water Heating* in the same series as this book). Sizing of a solar thermal system is critical if it is to do the job you want it to. In most cases, this job is to meet all or most domestic hot water requirements outside of winter and to augment the boiler's heating (for hot water provision) during sunny winter periods. In some parts of Europe, where sunny winter days are frequent and building heating demands are relatively low (in well-insulated buildings), solar collectors are increasingly being designed and sized to make significant contributions to space heating.

A common solar sizing rule of thumb is 2m² (21.53ft²) for each of the two main occupants, plus between 0.5m² (5.38ft²) and 1.2m² (12.92ft²) for every additional person, depending on how sunny your region is.

Most solar collectors will feed into a twin-coiled domestic cylinder or specially designed solar buffer tank. It is important to size these correctly according to the manufacturers' specifications. A small house might only require a 250-litre (55-gallon) store. Medium-sized houses may require a cylinder of around 400 litres (105 gallons). Larger properties will need one with a capacity for over 500 litres (132 gallons).

Fuel

There are two critical factors here: availability and quality. There are few corners of North America or Europe where wood pellets are unavailable these days, but it's important to check. Pellet quality is essential for the smooth and efficient running of any wood pellet stove or boiler. This should be discussed with the supplier or contractor at the planning and design stage to establish contact with your closest good-quality pellet supplier. Poor pellets have caused many problems with otherwise good installations.

Fuel reception

Once you've got a wood pellet fuel supplier on board, you will need to make sure that they can actually physically access your pellet store or silo. This will depend largely on the size of their bulk delivery lorry, the mechanism for filling (which ideally is by blower direct from the delivery truck) and the layout for access into and parking beside the property's pellet store. It's obviously much better to work out the entire process in advance, from delivery access to pellet filling system, rather than risk failure when the pellet lorry arrives at the start of the system's first winter. Pellet delivery trucks prefer to pump 20m (65 feet) or less from lorry to store.

Fuel storage and store sizing

The type of pellet store that's right for your premises depends on the availability of appropriate internal space, boiler location and so on, as discussed in the previous chapter. Pumping or other delivery to the store needs to be thought through carefully. The same goes for the pellet delivery mechanism for automated conveyance systems; access for auger or vacuum tube (which is much easier whatever the type of property construction between store and boiler room) is vital.

Ventilation is very important for health and safety (to reduce gas and fire risks) and to maintain the structural integrity of any pellet, more so if the pellets are to be blown in. This can be an open window or breather vent, but should be at least 170cm² (26 inches²), preferably larger. The store's foundations should be capable of bearing the maximum tonnage of pellets it can possibly hold. Self-standing flexible tanks or silos are often the best option, but they too need to be structurally up to the weights and forces involved, as well as having adequate footings.

Pellet stores are sometimes sized for the entire year or winter's supply. A rule of thumb for pellet use in a well-insulated property in the northern USA or northern Europe is about 400kg (881 pounds) for every 1kW of heat load (or boiler capacity); so a 10kW boiler might use about 4000kg (8818 pounds). In the UK, with a relatively mild climate, the formula would be more accurate with a figure of 300kg (661 pounds) of pellets per kilowatt, meaning that a 10kW pellet boiler could use as little as 3000kg (6613 pounds) per year. More accurate calculations are relatively simple:

- establish the kWh/yr predicted use (boiler capacity × hours of use/yr): e.g. 40,000kWh/yr, based on a 20kW boiler for 200 days at 10 hours a day
- work out primary demand (kWh/yr divided by percentage efficiency factor): e.g. 40,000 ÷ 0.80 = 50,000kWh/yr (0.8 being 80 per cent)
- calculate total annual fuel requirement (primary demand divided by energy content, assuming 4800kWh per ton of pellets): e.g. 50,000 ÷ 4,800 = 10,420kg (22,972 pounds, or 10.42 tons).

If you have a 20kW (or any size) pellet boiler and enough space to store 10 tons of pellets, then savings can be made by delivering them simultaneously. Most properties, however, simply don't have the space for this amount or cannot justify the expenditure on large additional structures. This latter type of property can determine their preferred pellet store size (say 1 or 2 tons) and then organize three to seven deliveries during the year.

Note that the walls and doors of a pellet store room should be fire-resistant and that they may need to meet safety regulations in your region (see section below). If the pellet store is in the boiler room, it is advisable to install an external boiler emergency stop switch outside of but near to the boiler room.

Back-up boiler requirement

If the heating load pattern is highly variable, it may be that a two-boiler (or back-up boiler) system makes economic sense. The most common approach would be to design the new pellet boiler to meet either peak load or base load. In peak load design, a single boiler system is able to meet the maximum imaginable heat load for the property. In base load design, a main boiler meets only the base load under common conditions. Peak load costs more in capital terms because the systems are bigger, but it maximizes the use of pellets. With peak load design, a boiler will quite often operate at a loading well below its stated capacity and both efficiency and emissions may be worse than necessary.

Base load systems are designed to meet between 60 per cent and 80 per cent of the property's annual heat demand. Generally speaking, a boiler with a capacity of around 60 per cent of peak load will do 80 per cent to 90 per cent of the work. A heat load duration curve based on historic fuel consumption data would help determine the best size. This kind of system needs a back-up boiler (for example, gas- or oil-fired). Capital costs and fuel use can be kept relatively low if there's an existing boiler to act as back-up. Alternatively, a heat storage silo could be used to help deliver the full peak load when required. Pellet boilers can be installed alongside, or to replace, existing gas, log, coal or oil boilers.

Peak loading may be the best option where user behaviour is such that the building will be in constant use or there is an additional and continuous process heat requirement. If there is an existing boiler (for example, gas-fired) with a good turndown ratio, it would be simple enough to keep this as back-up for meeting peaks and troughs in heat and hot water demand.

Health and safety

With pellet systems – from delivery and storage room to boiler and pressurized hot water delivery systems – there are a range of different health and safety issues. These will vary from one region of the world to another, according to local regulations. We cover them in more detail below.

Regulations and codes

Many of the regulations and codes that need to be considered concern health and safety issues, such as correct design and installation of the flue, pressure valves and so on. We cover these below.

Grants and payback

Grant schemes for biomass boilers come and go, but at the time of writing, in some regions of the European Union there are capital grants ranging from 15 per cent to 50 per cent available for the installation of clean biomass stoves and boilers, including wood pellet technology. In Missouri, USA, the Department of Conservation, in cooperation with the Department of Agriculture's Forest Service, is offering grants of up to $970,000 each for 'Fuels for Schools' projects involving turn-key wood-fuelled boiler installations funded by the American Recovery and Reinvestment Act (www.fuelsforschools.info).

In several US states, the Fuels for Schools programme promotes use of wood products as boiler fuel instead of fossil fuels. In Canada, energy efficiency grants are available from the Canadian Natural Resources Institute (CNRI) through the EnerGuide for Houses scheme.

Checking with your national forestry authorities, regional development organizations or energy and carbon advisory bodies or trusts can cut your final outlay considerably, reducing the payback period by years. In the domestic arena, much like the family car, boilers are rarely bought with payback in mind; more commonly the logic is one of reducing on fuel costs and carbon emissions.

Boiler sizing

Choosing the boiler type and size are probably the most important decisions in the design process. There has been a tendency to oversize boilers in recent decades in a crude attempt to ensure that they will always provide enough heat. This may well be a consequence of the low oil and gas prices experienced during the late 20th century. With money and fuel resources more precious than ever, it's important to get this figure right. Oversized heating equipment can create unwanted temperature swings in the property.

The capacity on the name-plate of an existing boiler is not good enough to determine the size of a new boiler. Historically, boilers were much larger, not least because of the cheaper fuel and less airtight and thermally wrapped buildings. Also, rules of thumb based solely on the size of your home or using a form that accounts for more factors might be good enough for an initial estimate, but should not be depended on alone for sizing the system. As the boxed examples illustrate, however, rules of thumb are a simplistic analysis based usually on rough figures. As such, they can generate quite a wide range of results. The matter is complicated even further by the fact that most pellet boiler manufacturers recommend using boilers significantly smaller (for example, 12–15kW) than the 20kW suggested by Table 5.4 later in this chapter.

Box 5.2 identifies ways of finding approximate boiler sizes given very basic information on a property (for example, floor area in square metres). For more accurate analysis, see the section on manual method later in this chapter, or consult the Earthscan publication, *Planning and Designing Bioenergy Systems* (see Sources of Further Information section at the back of this book).

Boiler system controls

Designed for the user to maximize comfort and/or minimize energy use, a good control system will give you heat and hot water only where and when you want it. This works with a timer plus individual room temperature controls, sometimes with zoning control for different areas of the property. Selecting the control options required can be achieved simply by studying the options provided by the various different wood pellet boiler suppliers contacted for initial advice on sizing and boiler selection. If you have a solar thermal collector, the control's software should be capable of managing this effectively alongside the heating circuit, hot water storage system and any mixer valves or pump stations in the system.

Box 5.2 Rules of thumb

Although rules of thumb are useful, particularly when gained from experience of designing and installing a specific boiler technology, they are not enough on their own. They can still be helpful in the initial stages, in preparation for early discussions with boiler manufacturers, suppliers or installers. Some rules of thumb include:

- Based on the total building footprint (ground floor area), the boiler output should be 1–1.5kW per 10m² (107ft²) of floor space (although this is very crude, and assumes good insulation levels, average ceiling heights plus temperate weather conditions). Assuming a ground floor area of 96m² (1033ft²), the formula would suggest a boiler size between 10kW and 15kW.
- Based on building volume (ground floor area only multiplied by height to start of roof space e.g. 96m² (1033ft²) x 5m (16.4ft) = 480m³ (16,951ft³)), the boiler output can be estimated by multiplying by a factor of 40, which in this case would be 19,200W or 19kW (although this assumes very low insulation levels and could almost certainly be downsized by up to 25 per cent, depending on insulation levels and air loss factors). In this way, the boiler specified might be in the range of 14kW to 17kW.
- According to the US DOE, a stove rated at 60,000BTU can heat a 186m² (2002ft²) home, while a stove rated at 42,000BTU can heat a 121m² (1302ft²) space (US Department of Energy – see Sources of Further Information section at the back of this book).

Phase 2 – specification

Specifying boiler appliance

If the boiler appliance is yet to be specified, once the heat demand and prospective boiler capacity has been identified, it's time to contact several suppliers by phone or email for technical details of the boilers they work with and any available performance reports or case studies. The supplier should be asked to size the system independently and share their calculations and assumptions with the potential customer. Wood pellet boilers do not represent a small expenditure, so good technical service should be expected before and after sales. With a few different quotes and specifications, a buyer can make an informed choice based on suitability for purpose and cost.

There are levels and sophistication of controls available in the wood pellet boiler market to suit all needs and tastes (see Table 5.2). This is much less true for the pellet stove market. Most standard boiler controls offer the ability to manage multiple heating spaces and several different heating zones (the latter with programmable timers and variable temperatures as well as continuous hot water).

As mentioned in Chapter 4, a good turndown ratio means that a boiler is relatively good at modulating and therefore more efficient and economic. In most situations, a boiler with a big turndown ratio (for example, 1:10) will be better than one with a smaller ratio (for example, 1:4).

At specifying stage, the actual details of operation and maintenance of the system should be investigated. This is usually possible by getting hold of a copy of the boiler's user manual or obtaining details from the supplier.

When ordering, the boiler manufacturer, model number and capacity should be specified along with where it is to be fitted. If the heat distribution system is not already in place, the same supplier or contractor may be able to quote or advise, although in most cases, this aspect of the system can be done adequately by a competent plumber as long as they familiarize themselves with the boiler function and overall system. The fuel-feed device and flue design are, wherever possible, best if supplied by the same manufacturer as the boiler to ensure optimum compatibility and a tried and tested system.

Boiler location and footprint

Most wood pellet boilers are designed to be housed in boiler rooms. Most pellet boilers are not kitchen appliances in the same way that a gas combi boiler is. In general, pellet boilers are typically installed in a shed or outhouse. Average boiler size is around 113cm (44 inches) wide and 109cm (42 inches) tall. The size of a normal domestic boiler room will ideally be 2m 6.5 × 6.5 by 2m (feet), although it is possible in a slightly smaller area. The actual size required can be worked out by scale drawing using the manufacturer's specifications.

Fuel feed

Most fuel feeds are either of the vacuum suction or auger types. At design stage, it's important to consider future access in case of blockages or faults in the ducts or mechanism. Relevant considerations for selecting these methods are given below.

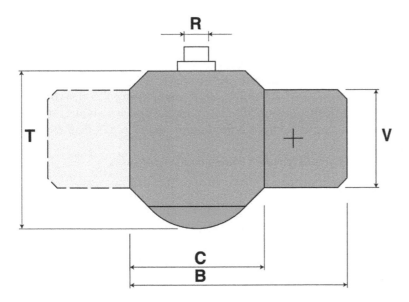

Figure 5.1 *Footprint of a PELLEMATIC boiler*

Source: ÖkoFen

Vacuum suction:

- avoid too many bends or turns in the filling and delivery pipes (otherwise blockages can occur);
- particularly appropriate for situations where the storage silo is in an adjacent or separate outside building;
- maximum length of 20m (65 feet).

Table 5.2 *Pellematic boiler data*

Boiler data:			PE08	PE10	PE15	PE20	PE25	PE32
Boiler – nominal capacity		kW	8	10	15	20	25	32
Breadth – completely	B	mm	1013	1130	1130	1130	1186	1186
Breadth – boiler	C	mm	645	700	700	700	756	756
Height – boiler	H	mm	1066	1090	1090	1090	1290	1290
Height – vacuum system execution	D	mm	–	1520	1520	1520	1710	1710
Depth – boiler	T	mm	691	814	814	814	870	870
Depth – burner cladding	V	mm	430	508	508	508	508	508
Forward and return – dimension		Inches	1	1	1	1	1¼	1¼
Forward and return – height of connection	A	mm	896	905	905	905	1110	1110
Smoke tube – diameter	R	mm	130	130	130	130	150	150
Smoke tube – height of connection	E	mm	664	645	645	645	800	800

Auger screw conveyance:

- the main disadvantage is that the long and inflexible auger screw means the boiler must be close and adjacent to the silo;
- auger mechanisms can go through walls between store and boiler room, but it can be difficult to angle a long and rigidly straight device like this through the wall (unless perfectly positioned in line with the boiler fuel inlet, it often necessitates making a much larger hole than the auger pipe's diameter just to fit it through);
- auger screws are generally cheaper and generate less noise.

Buffer tanks (or accumulators)

There are no strict rules about when and when not to use a buffer tank with a wood pellet boiler system. If there is ample space in the property and the budget permits, it is a good idea for the smoothing effect it has on delivery of heat and hot water. The decision is also likely to be affected by the boiler control and combustion design properties, so the manufacturer or supplier is likely to have a view. Boilers that modulate well should not really need a buffer tank, but it is possible to identify a few scenarios in which they can be an important element:

- when very short cycles of heat are needed and to reduce modulation and cycling when the building has a very low heat load (e.g. an eco-house);
- to facilitate highly stabilized combustion with minimal modulation and cycling;
- with a large solar thermal input (common in Austria and Germany) or a well-engineered heat exchanger for the solar input into the accumulator (e.g. stratified type);
- where there is more than one boiler in the system serving the same heat load.

When a buffer tank (accumulator) is selected for a system, it has to be sized first and then sited as conveniently as possible for the corresponding pipework. This gets slightly more complicated when there is a secondary thermal input to space heating (such as from solar thermal collectors). A general rule of thumb for sizing is that a buffer tank should hold at least 25 litres (6 gallons) per kilowatt of boiler output (for example, around 450 litres (118 gallons) for an 18kW boiler).

System configuration

Installation costs and operational efficiency can be optimized by choice of system configuration, the design and length of pipe runs, heat distribution circuits and connection types. Again, thinking about future access to important elements in the system is good at this phase. Simple sealed systems are the most commonly used for pellet boilers.

Hot water radiators can be made more efficient by fitting separate zone controls for different areas of a larger property (particularly effective if large areas are not used much). Automatic valves should be used on radiators, controlled by thermostats in each part of the building.

Flue gas system

Whether installing from new or renovating an existing chimney, it is vital that the flue is resistant to corrosion. Correct selection and sizing of the flue is very important for efficient functioning of the system. The cross-sectional area of flue required depends on factors such as the nominal heating rate and the height of the flue. Check on local codes and regulations (see below) and ask the flue manufacturer for their calculations and recommendations. Table 5.3 provides some rules of thumb, but this is a detail to check with local regulatory authorities.

Automatic cleaning and de-ashing

This is usually a matter of personal preference, depending on the installation scenario and user-operators involved. It's something to bear in mind when specifying the boiler type and desired sophistication of automation. One thing worth noting is that good-quality pellets leave only about 1 per cent ash, so if a boiler is consuming 10 tons a year it will produce around 100kg (220 pounds) of ash annually.

Manual method

The following is a manual method that can be used for initial calculations and as a basis for discussions with manufacturers and suppliers – it is limited by the fact that many assumptions are made regarding the heat requirements of the building, the local climate, building insulation and heat distribution system, and it is carried out before any particular boiler has been chosen. (For more detailed design information, readers may wish to consult the Earthscan publication, *Planning and Designing Bioenergy Systems* – see Sources of Further Information section at the back of this book.)

The worked example that follows is an attempt to assess approximate boiler size for a detached rural house in the UK. There are no extreme weather conditions and the temperature rarely drops below –3°C (27°F). Preferred internal temperatures are 21°C (70°F).

This is a two-storey house 12m (39 feet) long by 8m (26 feet) wide, giving a floor area of 96m² (1033ft²). The wall area is 200m² (2152ft²) (two walls, 12m (39 feet) by 5m (16 feet) high to start of roof space = 120m² (1292ft²), plus two walls of 8m (26 feet) by 5m (16 feet) = 80m² (861ft²)).

There are 16 windows with a total surface area of 40m² (431ft²): eight downstairs (each measuring 2m (6.5 feet) by 1.8m (6 feet)) and another eight upstairs (each measuring 2m (6.5 feet) by 1m (3.3 feet)).

In this house, as in most cases, the roof (or ceiling) area (96m² (1033ft²)) is the same as the floor area.

For this example, we have selected typical U-values for the floors, walls, ceiling/roof space and windows. Similarly, the ventilation factor selected is fairly typical of a standard UK house.

Table 5.3 *Flue sizing chart rules of thumb*

Nominal heating rating	Chimney cross-section size
Up to 10kW	14cm
10–20kW	14–16cm
20–30kW	16cm

By working through the stages of Table 5.4 making manual calculations, it is simple enough to arrive at an approximate heat loss figure for the property. Then, by adding an estimate for domestic hot water demand (we have assumed 250W per person living in the property), the resulting figure is the required boiler size. For this example, we have used a simplistic approach; in practice, however, one would attempt to account for the number of hot water applications in the property, as well as the number of inhabitants. Generally speaking, one tap or shower requires around 20kW instantaneously for a limited duration.

To calculate the annual heating demand, simply multiply the boiler size (in kW) by the peak-load-equivalent hours (which are generally estimated as 1200 hours a year for UK climatic conditions): 24 × 1200 = 28,800kWh.

To calculate the yearly fuel requirement, divide the annual heating demand by the rated efficiency of the selected boiler (in this case, we assume boiler efficiency of 80 per cent; while manufacturers offer design efficiencies of around 90 per cent, the actual annual average efficiency – including start-up and shut-down operations – is normally between 75 per cent and 85 per cent): 28,800 ÷ 0.80 = 36,000kWh/yr. Then divide this figure by the energy value of the fuel (quality wood pellets are calculated at 4800kWh/ton) to obtain a tonnage for the annual fuel requirement: 36,000 ÷ 4800 = 7–8 tons of pellets a year.

A worked example

Manual forms such as Table 5.4 generally result in a generous figure for heat loss. The worked example above, for instance, indicates a heat loss figure of 19,152W. We have then added a token figure for the expected hot water demand, which brings the suggested boiler size up to 20.2kW. Some boiler manufacturers or suppliers may not calculate the domestic hot water demand because the loading of the cylinder can be timed to coincide with the heating programme. Also, many households generate some heat from cookers, fridges, washing machines, lights and other domestic appliances, which supports the boiler plant.

One question to think about is under what circumstances is the constant heat load going to be, say, 20kW and how long in each timed period will the boiler be enabled? If it is roughly half the time, then arguably a 12–15kW boiler would be sufficiently big. Using standard manual or even online forms for calculating heat loss and boiler size, it is important to double check your resulting figures against those recommended by your boiler supplier.

Using software

Clever software packages can be obtained, often free or on a free trial basis, from a variety of sources in Europe, the USA and Canada (some of the better options are listed in the Sources of Further Information section at the back of this book). Many of these work very well and a few are specific to biomass, or at least have an option for wood pellet technology. Most are available online or as downloads from relevant websites.

Table 5.4 *Basic heat loss and boiler sizing form for manual calculation*

		A	B	C	D	
	Dimension	Size	U-value (except row 5, which is ventilation factor)	Temperature differential between expected outside lowest and preferred inside temperatures (°C)	Heat loss for property (W)	
1	Floor area (m²)	96	1	24	2304	Multiply floor area (1A) by U-value (1B) then multiply the result by the temperature differential to get heat loss of floors (here we assume −3°C outside and 21°C internal preferred temperature)
2	Wall surface area (m²)	200	1.68	24	8064	Multiply wall surface area (2A) by the U-value (2B) then multiply the result by the temperature differential to get the heat loss from the walls
3	Windows surface area (m²)	40	2.8	24	2688	Multiply window surface area (3A) by the U-value (3B) then multiply the result by the temperature differential to get the heat loss from the windows
4	Roof area (ceiling area m²)	96	0.6	24	1382	Multiply ceiling surface area (4A) by the U-value (4B) then multiply the result by the temperature differential to get the heat loss from the windows
5	Building ventilation or air loss (5A should be the building volume m³)	480	0.33*	24	3802	Building ventilation loss = 5A (calculated as building height to start of roof multiplied by building floor area) multiplied by 0.33 multiplied by average temperature differential
6				INITIAL BUILDING HEAT LOSS ESTIMATE	18,240	
7	Compensate for extreme weather exposure conditions (if any) by adding roughly 5–15%				19,152	For this example of detached rural hill-based house we have added 5%
8	Expected domestic hot water demand (W)				1000	Assuming 250W per occupant (with 4 people living here) = 1000W for domestic hot water demands
9	REQUIRED BOILER CAPACITY (W)				20,152	
	BOILER CAPACITY (kW)				20	Divide 9D by 1000 to convert from watts into kilowatts

In ideal circumstances, the results of a manual assessment can be checked against one or two of the available software options before discussing requirements with potential boiler suppliers.

Project development/community and multiple-residence projects

Table 5.4 is designed to help project development for new wood pellet or other boiler projects. The same logic applies to community projects as to domestic or commercial installations. With community initiatives, however, there are a number of additional issues that need to be addressed:

- model for developing the project (e.g. a developer initiative with community investment and benefit; a community-led initiative with bought-in expertise);
- typical barriers to community projects (such as the need for a determined core development group, access to finance, access to specialist knowledge);
- legal structures (someone or some organization has to own the plant and be ultimately responsible for it – there are various models appropriate for community ownership, including a company limited by shares, a company limited by guarantee, an industrial and provident society or a registered charity).

Building engineering issues

The main building-related factors that need to be considered at the design stage are:

- space available for locating boiler, pellet store and, if required, buffer tank/accumulator;
- space available for solar collector and related pipework, if required;
- flue design (whether new or refurbished);
- fuel store position in relation to boiler fuel-feed inlet;
- foundations and footings for the wood pellet store;
- sufficient access for blower lorry delivery.

Building control, air pollution regulations, safety and fire prevention

Regulatory control is common for medium and large heating systems in most countries. However, the amount and nature of control varies greatly. In the USA there is minimum regulation, with legal processes sorting out any issues that may arise over performance or safety. In the UK, the scenario is one of high regulatory intervention. This is true for both building and air pollution control.

Table 5.5 *Stages of project development*

Development phase	Main tasks involved
Define need/objectives	*What do you want done and why? Identify specific levels of comfort and automation required from new boiler system. What is the expected result?*
Technical assessment	*Work out the heat load and hot water requirements to establish an initial estimate of required boiler capacity. Consider other aspects of the design: heating circuit designs, energy efficiency measures, level of automation required, solar input, buffer tank or not, etc.*
Financial assessment	*Make initial estimates of boiler and system costs. Analyse likely fuel consumption and costs for comparison with alternatives. Check out availability of capital grants*
Planning control and regulatory issues	*Identify any planning control, safety or other regulatory issues. Talk to the authorities involved where documentation or permission may be required*
Begin procurement	*Decide which suppliers to contact, then get in touch for the relevant technical data and advice (including a full heat load assessment)*
System selection	*Sift through the alternatives to find the boiler and system that best suits your objectives and budget, making sure that the supplier has the expertise and experience to supervise the installation as well as stocking the most appropriate boiler equipment*
Specification	*Agree detailed specification and delivery/installation dates with selected supplier. Sign contract when completely satisfied with offer*
Delivery and installation schedule	*Draw up a written delivery schedule with supplier and installer, ensuring that equipment is delivered on time and in the right order. This schedule can then be used to devise a definitive and realistic installation schedule with the agreement of all parties*
Signing off	*When the system is fully installed and has been fired up for several test runs, the installation phase can be signed off. At this stage it is important to ensure that you are happy with the instructions for operation and maintenance and that a plan for servicing the boiler system is designed*

Building control

Given the importance afforded to the regulation of heating systems in the UK, it provides an adequate model for looking at the main issues. For details on recommended and essential codes and regulations in the USA, see the Building Energy Codes Program's website. The principal source for details on British building controls for wood pellet stoves and boilers is the UK government's online Planning Portal, Part J of which deals with all wood fuel as if it were the same as any other solid fuel. Other detailed issues relevant for consideration and research at the design stage are covered by:

- Part A – structural and building issues;
- Part B – fire safety;
- Part F – ventilation (this is important structurally for pellet stores that are filled by blowing and also to protect the health of occupants);

- Part G – storage system design, valves and safety devices, discharge of any excess water;
- Part L – energy conservation;
- Part P – electrical safety.

More detail on this is provided in the Sources of Further Information section at the back of this book.

Pressure systems safety regulations

As with any boiler system, a serious flaw in design or problem in operation could cause a fire, steam-burst or even an explosion that could kill or injure persons in the building or boiler room. Bad design, installation or shoddy maintenance can contribute to such a problem. Particularly for larger systems, operator error or poor training can be factors in this. Most countries have regulations or national guidance on safety for pressurized systems. The EU Pressure Systems Safety Regulations (PSSR) state that pressure systems should have:

- a written scheme of examination (WSE) in place that has been certified by a competent person;
- be subject to inspection by a competent person in accordance with the WSE;
- a body that has the competence to undertake these functions, which it is the responsibility of the user or owner to appoint.

The Canadian Environmental Management Act (1994) provides the basis for regulations concerning domestic solid fuel-burning appliances, including pellet systems. In the UK, a competent person for solid fuel heating systems is defined by the Heating Equipment Testing and Approval Scheme (HETAS). Wherever in the world you are, however, most building owners will employ a contractor or installer who can provide adequate and competent cover for this aspect of the design. The single most important thing to ensure is that the expansion vessel in a sealed system is large enough to take the system's maximum potential increase in volume.

Health and safety in the workplace

If a heating system is for a building where people are employed, such as a school or offices, then standard national health and safety regulations will apply. For regional contacts and more details on this issue, see the Sources of Further Information section at the back of this book.

Air pollution control

In the USA, the Environmental Protection Agency provides standards for air quality. For the UK, any heating installation between 400kW and 3MW is required by the local authority to have an air pollution control permit. UK wood pellet heating schemes can theoretically operate at a capacity of up to

20MW without an air pollution control permit if they can prove that the source material for the pellets is 100 per cent clean biomass.

Heating systems any larger than 3MW in the UK are regulated by the Environment Agency, which is responsible for implementing the Pollution Prevention and Control Act (1999) and is likely to insist on emissions monitoring (either continuous or periodic). Pollution control regulations are, of course, subject to change, particularly during times such as these with increasing international concern about climate change. At the time of writing, however, smaller systems in the UK (400kW or under) are not required to obtain any air pollution permit – except that, technically speaking, even a wood pellet stove or boiler is in danger of breaching smoke control area legislation, which covers most of the UK's urban areas. Some stoves and equipment are exempt, but very little pellet combustion technology has been registered to date. That which is registered can be used in smoke control areas as long as the fuel 'does not contain halogenated organic compounds or heavy metals as a result of treatment with wood preservatives or coatings', the legislation explains.

Fire prevention

Fitting appropriate protective devices is essential. Electronic sensors, safety valves and fuel cut-off mechanisms are all very important for maintaining safety if temperatures or pressures exceed allowable limits. The safety devices should be fail-safe – that is, they will always turn the boiler off when a dangerous fault is detected. Boiler safety, operator and maintenance instructions should always be kept close to the boiler itself.

The main fire risks are back-burning in the fuel feed, overstacking or a blockage in the combustion zone. It cannot be overstated how important it is that the boiler system is designed and equipped to deal with and prevent anything like this happening. A safe boiler system usually depends most on sophisticated controls that are able to monitor everything from combustion gases and fuel availability to unusual variations in temperature or pressure. When selecting a boiler, full details of safety features should be a top priority for the potential buyer. A chain of multiple fail-safe mechanisms, including temperature sensors, airlocks, water sprinklers, foam extinguishers and fuel cut-offs mechanisms should be considered. Back-burning, particularly with an automated fuel-feed system, is the biggest single risk. European legislation is expected to stipulate a minimum of two protection levels against back-burning for pellet burners, but three levels of protection for the burner are already commonly used in combination: safety shut-off valves on the fuel feed, an internal water sprinkler system, and/or physical disconnection of the pellet supply.

References

ACCA (2006) *Manual J Residential Load Calculation*, ACCA, Arlington VA
CIBSE *(2007) Domestic Heating – Design Guide*, CIBSE, London
CNRI website, www.oee.nrcan.gc.ca/publications/houses/M144-13-2005-1E.cfm (accessed 21 November 2009)
Deutsche Gesellschaft Fur Sonnenenergie (DGS) and Ecofys (2004) *Planning and Installing Bioenergy Systems – A Guide for Installers, Architects and Engineers*, Earthscan, London

Government of Canada *(1994), Environmental Management Act,* www.bclaws.ca/Recon/document/
    freeside/--%20E%20--/Environmental%20Management%20Act%20%20SBC%202003%20
    %20c.%2053/05_Regulations/44_302_94.xml (accessed 23 November 2009)

Laughton, C. *(2010) Solar Domestic Water Heating – The Earthscan Expert Handbook for Planning,*
    Design and Installation, Earthscan, London

Planning Portal website, www.planningportal.gov.uk/england/professionals/en/4000000000503.
    html (accessed 23 November 2009)

US Department of Energy, Building Energy Codes Program website, www.energycodes.gov
    (accessed 23 November 2009)

US Environmental Protection Agency, Air Quality Planning & Standards website, www.epa.gov/air/
    oaqps/index.html (accessed 23 November 2009)

US Fuels for Schools website, www.fuelsforschools.info (accessed 21 November 2009)

6
Installing Wood Pellet Technology

The installation of a pellet heating system involves a wide range of skills and specialisms. Property owners new to wood pellet technology, and probably most of those already familiar, are likely to want to pay the equipment supplier to install and commission the complete system, and they themselves will only need a limited understanding of the installation process. However, a few may have the necessary skills to carry out some of the installation process themselves, or access to personnel with those skills.

This chapter runs through the various elements of the process so that anyone considering an installation will know what to expect and what may be possible. A few pointers are given where they might help first-time installers to avoid certain problems.

Most of the chapter relates to pellet boiler systems, with a brief discussion of pellet stoves.

Elements of the installation process

There are a few general points worth mentioning relating to the boiler itself and the space to house it. Pellet boilers need to be housed in a boiler room or equivalent that is strong enough, large enough, sufficiently ventilated and appropriately fireproofed. Exact requirements will vary from boiler to boiler and from country to country, but the following points are worth remembering, particularly in that they may differ from an equivalent fossil-fuelled boiler:

• Pellet boilers are heavy. Even a small domestic boiler may be over 200kg (440 pounds), and the complete installation with filled accumulator tank can easily be over 1 ton, not including the fuel store. This has implications for the floor slab, but it also has implications for receiving and locating the boiler itself, especially if it is to be sited in an existing building or even basement. You will need a strategy in place for manoeuvring the boiler before it arrives.
• Pellet boilers require ash removal, cleaning and other regular maintenance activities that may take up significant space around the boiler. The space required for controls, pipework, fuel conveyors, accumulator tanks and so on may also be greater than you expect.

- Pellet boilers are solid fuel appliances, containing significant amounts of burning material during operation. The regulations regarding fireproofing may be significantly more stringent than for a fossil-fuelled boiler.
- Pellet boilers do not have balanced flues. They require adequate ventilation of the boiler room. Specific requirements will vary but, as a rough guide, you should expect a permanently open ventilator at least as large in diameter as the boiler flue pipe (the flue itself is covered later on).

Fuel store

The fuel store size has been discussed in Chapter 5, but is likely to contain several tons of pellets, and the structure needs to be built with this in mind. The fuel store may be built from scratch *in situ* or, as with a FlexiTank, supplied as a kit. In either case, the boiler or tank supplier's specifications should be followed closely.

Specific requirements will include structural strength, weather protection and ventilation. Ventilation is necessary to reduce the risk of explosion, and to avoid damage during pneumatic fuel-feeding. The fuel store will usually be in a separate room to the boiler, with a fireproof wall in between, although some countries do not require this for smaller systems. There should be at least two systems to prevent burn-back to the main store.

Clearly, the store needs to be sited appropriately for both pellet delivery and feeding to the boiler. This positioning will be particularly critical where auger feeds are used. It may be easier to install or construct the fuel store first and then align the boiler to meet the auger before finally fixing everything in place.

The main auger connecting the fuel store to the boiler will often pass through the intervening wall at an unusual angle. It is often necessary to make a large hole in this wall to allow the auger to manoeuvre. Once all elements are secured, the remaining gaps can be filled in. If the boiler room and/or fuel store are newly constructed it may be easier to build or complete the intervening wall after installation of the auger.

Flue

The importance of a well-designed, well-built and well-maintained flue cannot be overstressed. If the flue fails to perform adequately in all circumstances (including failure of the power supply) then lethal gases could build up in the boiler room and make their way into adjoining areas.

The boiler supplier will provide information on what they require in the way of a flue, but there may well be more stringent requirements in the local regulations, often with good reason. It is important to find out what is required and comply with it. A properly compliant installation may end up more expensive than you anticipated, but it will be legal, easier to insure, easier to sell when you move on and it could save your life.

If you are using an existing masonry chimney, it will probably require lining and insulating to maximize the natural draught and to avoid condensation of tars as well as leakage. This can often be achieved by inserting a single-skin liner and filling the remaining cavity with insulating material.

If installing a new flue pipe, this will probably be of insulated double-skin stainless steel (similar to that in Figure 6.1). Your boiler supplier should be able to advise on local requirements, including the necessary height, diameter, angles of runs and configuration of bends. Generally, bends should be kept to a minimum, as should horizontal runs (if permitted at all), the diameter should never decrease, the top of the flue should be significantly above the eaves of the roof it passes through or is adjacent to, there should be a cowl or similar at the top to discourage downdraughts, and the outlet should be some distance away from opening windows or other ventilation intakes.

If there is an existing flue from a gas or oil boiler, then bear in mind that an equivalent-sized pellet boiler will produce a greater volume of flue gas – typically around 40 per cent more. So a factor of 1.4 should be applied if using flue-sizing tables based on fossil fuels to check the chimney capacity.

A properly installed, good-quality pellet boiler burning good-quality pellets will give very good combustion, and clean combustion gases with very low particulate content. However, some local regulations may require that a cyclone or precipitator is included in the flue system to deal with any particulates.

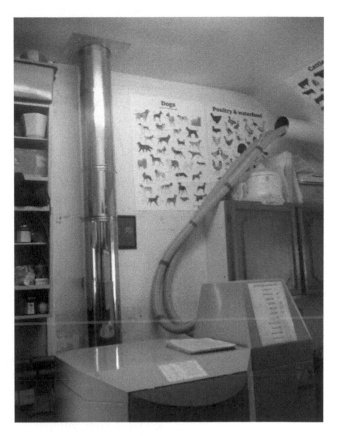

Figure 6.1 *Installed stainless steel flue pipe with adjacent vacuum suction tube*

In small systems, the boiler supplier may not have allowed for this local requirement, in which case it will be down to the installer to add such an element to the flue output of the boiler.

Plumbing

The plumbing aspect of a simple pellet heating system is very similar to that of a conventional boiler system. The heat distribution and hot water systems for the property are likely to be identical, with the pellet system connected in place of a conventional boiler.

However, the pellet heating system may also be more complex – for example, it may include an accumulator. Your supplier should give guidance on how the accumulator and boiler ought to be connected and you can often then treat the combined boiler/accumulator system as though it were a boiler.

Systems with more than one boiler are more complicated still, with the range of plumbing options becoming greater with each additional piece of equipment. Larger heating systems also have a number of optional extras in the design, which all add complexity. The schematic layout will have been decided on at the design stage, but the job of creating the pipework system according to that schematic can become dauntingly complex for the inexperienced plumber.

Pellet boilers are rarely available as condensing units and rarely fitted with separate condensing economizers, as condensate from a wood burner is harder to deal with than condensate from a gas or heating oil boiler. A pellet boiler system is therefore more likely to include a bypass loop, to allow hot boiler output water to mix with the cold return and avoid the low return temperatures that lead to condensation. In small systems, this loop may be internal (i.e. inside the boiler) and will not require plumbing, but larger systems may require it as part of the pipework.

It is also fairly common to include a bypass loop after the accumulator. This is used to mix hot and cold water to the heating circuit flow, in order to avoid excessive temperature and distribution heat loss. This is known as flow compensation.

Sometimes the amount of cold water flowing through the compensation loop is automatically adjusted according to the outside temperature. This is known as weather compensation, and reduces heat losses by running the heating circuits at lower temperatures in mild weather.

Controls

The controls for a pellet boiler are very different from those of a conventional boiler, and are supplied as part of the boiler package. The control requirements for the heat distribution and hot water system are the same, but their integration with the boiler control system is not.

Using a conventional heating controller in parallel with a pellet boiler controller is unlikely to work unless the combination is approved by the boiler manufacturer. Generally speaking, the boiler will come with a fully integrated control unit that can be used to program heating and hot water requirements, including multiple areas of the building, as well as managing boiler operation itself.

In a packaged boiler system, the control system will also include the required safety systems. Technically, control systems and safety systems are separate as they have different functions. However, the two are effectively combined in the control unit. All the installer needs to do is follow the instructions carefully, bearing in mind that correct installation is necessary to ensure safety as well as performance.

Electrical connections

Pellet boiler systems require a significant amount of electrical connection work, including mains voltage connections for pumps, fans, motors and motorized valves, and generally low-voltage connections for controls and sensors. This work requires a mixture of standard electrician's skills to connect to the mains system legally and safely, and specialist guidance from the boiler supplier and/or system designer to connect the system correctly.

A competent and experienced electrician may still need help if unfamiliar with pellet systems, especially if working from a poorly translated installation manual. Power connections to items such as motors and fans should present no surprises and merely need to be connected in line with local legislation as per any other installation. However, the interconnections between the control unit or units and various sensors and devices are likely to be complex and unique to each system. Incorrect connection could lead to permanent damage as well as malfunction, and it is vital that the manufacturer's instructions are followed to the letter.

Commissioning

Pellet boilers are popular partly because they offer the most automated and least troublesome zero-carbon heating option for many situations. To achieve this, systems have complex controls and complex fuel-feed systems, over and above the usual complications of an automatic heating system.

These systems will only work correctly if the system is properly commissioned, and this is probably the part of the installation process that is most dependent on specialist input. Commissioning will be more effective and far quicker if carried out by someone who has done it before with this type of boiler. The boiler will probably come with instructions for commissioning the system, and certain competent people may be able to carry out the job effectively without specific training. However, using a qualified installer for commissioning the system (if nothing else) may be the most cost-effective solution, as well as potentially meeting legal and other requirements (see below). As with other stages of installation, there will always be aspects of commissioning that can be carried out by any competent person, such as filling the fuel store and filling and pressurizing the heat distribution circuit or circuits.

A number of manufacturers offer 'energy box' solutions, where the heating system, including boiler, accumulator, flue and fuel store, are delivered as a single preconnected containerized unit. Installation in this instance is limited to laying foundations and connection to services and heating systems. Commissioning can then be carried out by the supplier's representative, often on the same day as delivery.

Operation and maintenance

Day-to-day operation of the boiler and heating system should not present any problems once the control system is understood. It is the responsibility of the installer to provide the user with clear and comprehensive operating instructions for the whole system, so that it may be operated efficiently, safely and according to the terms of the warranties.

Apart from day-to-day operation and routine fuel deliveries, there are regular maintenance tasks to be carried out.

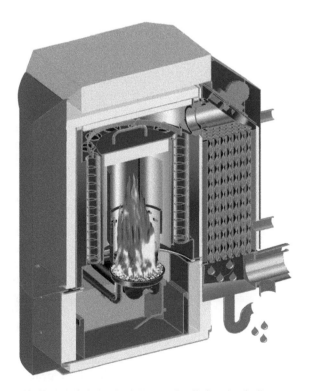

Figure 6.2 *Schematic cutaway of self-cleaning boiler*

Source: ÖkoFen

Ash removal will be required at intervals, depending on the boiler's ash collection system. The boiler may be described as having 'automatic de-ashing' but the ash still has to be emptied. The ash is generally removed from the boiler in a closed container, and the ash disposed of, often on the garden. If the ash has to be disposed of elsewhere, you will have to check the local waste disposal regulations or use an approved waste handler.

Remember that the ash may be hot when you remove it! Many pellet boilers are self-cleaning, in that the insides of the fire tubes are automatically scraped at intervals. However, in some boilers this is a manual task, and all boilers will require a full clean as part of their annual service. Regular servicing by approved personnel may be a requirement of the manufacturer's warranty.

The flue will also need sweeping once a year. In some countries, the chimney sweep will dismantle the flue, clean it thoroughly, reassemble it and provide a certificate to prove that the job has been done. In other countries, the sweep is not permitted to do this, and the flue design must allow for sweeping without dismantling.

Pressurized systems may require inspection at regular intervals, depending on the size and pressure of the system and the local legislation.

Pellet stoves

Installation of a pellet stove is clearly simpler than installing a boiler, as there is no plumbing to consider, no complex control system required and no automated fuel-feed system. The stove is also considerably lighter and easier to manoeuvre. The requirements for fireproofing, ventilation and flue are just as vital though, if not more so, because the stove will be operating in an occupied space. The first stage of installation is preparation of the room, including laying a fireproof hearth and, where necessary, installing or upgrading the flue. The stove can then be located on the hearth and the flue pipe connected. Electrical connections are made and fuel added, so the stove is then ready for its first lighting as per the manufacturer's instructions. If the stove has a back boiler, then all the recommendations relating to boilers come into play.

Skills required to install a wood pellet system

So who can install a pellet boiler system safely and legally? It is clear from the sections above that there is likely to be a lot of general building work involved, plus some plumbing and electrical tasks, perhaps carpentry, perhaps roof work, and some specialist input relating to the boiler itself and the total system layout. A large project is likely to involve a team of specialists to cover each aspect. Smaller projects are more likely to be carried out by an individual from one of these trades who has diversified sufficiently to cover some or all of the other aspects.

In principle, much of the work for a small installation could be carried out by a skilled DIY practitioner. However, in some countries there are restrictions on who is legally permitted to carry out certain aspects of the work. There are also often constraints on installation personnel from other sources:

- Capital grants towards the cost of the equipment may be dependent on the use of approved installers.
- Equipment manufacturers may only supply to their own network of trained installers, or only warranty equipment installed by them.
- Use of a certified installer may provide an easy route to approval of the installed system by the appropriate regulator.
- Insurance and mortgage providers may require documentation from certified installers.

In practical terms, even for a large system, much of the actual work could be carried out by a competent builder or plumber with limited support from the boiler supplier. However, some aspects require more specialist skills, and will usually entail detailed training of the individual in question. It may be practical in some circumstances for less specialist personnel to carry out the installation under the guidance of a certified installer who would check and commission the installation and provide any certification required.

Relevant codes and regulations

The regulation of pellet heating installations was covered to some extent in the previous chapter. However, there are often regulations relating specifically to the installation process and to who may carry it out.

In the UK, heating installations are covered by the Building Regulations, which are enforced by building control officers. However, these officers are responsible for monitoring the complete process of building construction and renovation, and cannot be expected to understand all the detailed requirements of all possible building services. The government deals with this by setting up 'competent person' schemes, and the scheme for solid fuel heating systems (including pellet boilers) is HETAS.

Members of HETAS are approved to install solid fuel heating systems. Thus if the system is installed by a HETAS-registered installer, then the system is considered to meet the regulations, and the installer can provide a certificate to say so. Unfortunately, the rules that HETAS installers follow were developed to cover coal boilers and they are not always appropriate for pellet systems. However, this issue is being addressed as pellet heating becomes more popular in the UK.

Most electrical work in the UK is carried out by qualified electricians, but there are no regulations to prevent a homeowner, for example, wiring in their own system. However, in theory at least, this work would then have to be approved by the building control officer.

Where a large system is being installed in a commercial situation, the full range of relevant regulations can be very large. The following list shows some of the more relevant pieces of legislation in the UK:

- The Construction (Design and Management) Regulations 2007
- The Pressure Systems Safety Regulations 2000
- The Regulatory Reform (Fire Safety) Order 2005
- The Electricity at Work Regulations 1989
- The Dangerous Substances and Explosive Atmosphere Regulations 2002

- The Confined Spaces Regulations 1997
- The Work at Height Regulations 2005.

Most developed countries have an equivalent collection of regulations.

In the USA, the installation process is generally less regulated. There is some national regulation from the Department of Housing and Urban Development (HUD), but most aspects are covered by local building codes. Generally these relate to the installation itself, rather than to who carries it out.

In mainland Europe, regulation again focuses on the installation. Germany and particularly Austria have high emission standards to be met by manufacturers. Also, many countries require an annual service and chimney clean by a qualified professional who provides the householder with a certificate. This emphasis on high-quality equipment and regular maintenance takes the place of the UK's emphasis on the detail of the installation, particularly in relation to the flue design.

Generally in Europe, the standard of an installation is regulated by the local or national building codes, which vary enormously in their ability to regulate pellet systems. However, it is possible for the client to specify that the installation must be European Conformity (CE) marked. It then becomes the responsibility of the installer or prime contractor to ensure that the whole installation is installed and commissioned to this standard and that the necessary documentation is produced. This is different from the CE marking of individual items of equipment, such as the boiler, where it is the manufacturer's responsibility to comply.

Trouble-shooting

If a problem occurs, the boiler controller should display error messages that, with the aid of the manufacturer's manual, will help the installer or maintenance technician identify and remedy the problem. Tables 6.1 and 6.2 are designed to cover some of the more basic problems that might arise, some of which could be remedied without the need to call in an expert. Table 6.1 covers boilers and Table 6.2 stoves.

Table 6.1 *Simple fault diagnosis for wood pellet boilers*

Fault	Some possible remedies
System does not switch on	Check power is connected. Check fuses.
No fuel reaching burner	Check fuel store is not empty. Check augers have been primed with fuel. Check boiler control is set to on and heating controller is demanding heat. Check augers are not blocked. Check auger drive motors are functioning/connected.
Fuel in burner is not burning	Check boiler control is set to on and heating controller is demanding heat. Check ignition system is connected. Repeat initial ignition procedure.
Fuel is burning but hot water is not being produced	Check pressure in heat distribution circuit is adequate. Check heating controller is demanding heat. Check heating pump is connected.

Table 6.2 *Simple fault diagnosis for wood pellet stoves*

Fault	Some possible remedies
System does not switch on	Check power is connected. Check fuses.
No fuel reaching burner	Check fuel store is not empty. Check auger has been primed with fuel. Check stove control is set to on. Check auger is not blocked.
Fuel in burner is not burning	Check control is set to on. Repeat initial ignition procedure.

Risk mitigation

Table 6.3 identifies some of the risks that may arise during and after installation, and suggests possible mitigation strategies. The centre column considers risk from the point of view of someone managing an installation process, either for themselves or for a client. The right-hand column is for those paying a supplier or installer to carry out the complete project. Some options may be more or less relevant depending on the scale and cost of the system.

Table 6.3 *Mitigating risks during wood pellet equipment installation*

Risk	Mitigation options (installer)	Mitigation options (client)
Commissioning date is delayed	Include contingency in schedule. Request firm delivery dates from suppliers as part of quote. Plan realistic critical path	Request firm completion date from installer/supplier as part of quote. For large projects, consider penalty clauses
Cost goes over budget	Include contingency in budget. Request fixed-price quotes for extended period from suppliers. Comprehensively check budget	Request fixed-price quote, not estimate. Check quotes to ensure full costs are included
Excessive teething troubles following commissioning	Ensure system designed by experienced professional. Ensure good after-sales care in suppliers' quotes. Allow for sufficient commissioning time in budget and schedule	Ensure good after-sales care in installer's quote. Choose installer/ supplier with good record
System fails to provide sufficient heat at the required times and places	Carry out comprehensive heat load assessment at an early stage. Ensure system is designed by experienced professional. Check heat distribution system and controls are adequate. Check commissioning and operating procedures	Carry out comprehensive heat load assessment at an early stage, or check that installer can and will do so. Choose installer/supplier with good record. Ask installer to check heat distribution system and that controls are adequate. Ask installer to check commissioning. Check operating procedures are correct
Boiler or other elements fail to work satisfactorily	Choose suppliers with good record. Ensure good warranties are provided at quote stage. Check fuel quality. Check commissioning and operating procedures	Choose installer/supplier with good record. Ensure good warranty is provided at quote stage. Check fuel quality. Ask installer to check commissioning. Check operating procedures are correct

7
Wood Pellet Manufacture

Source of the raw material

Wood pellets are usually manufactured using by-products from the timber industry, namely sawmills and joinery workshops, although other sources such as arboreal cuttings and sustainable forestry are also viable. The availability of particular sources varies between regions and countries; however, sawmills continue to be the primary supplier of raw materials for the pellet industry. Ideally, raw materials should be sourced locally to avoid the additional cost and carbon emissions associated with transport over long distances. While pellets are cheaper to transport per kilowatt of user output than wood chips, this is not the case when compared with oil or gas delivery. There are examples where the radius for local pellet supply is over 500km (310 miles), but this is not ideal. In 2006, the demand for pellets increased enormously as a result of rising oil prices. At this time, problems in the supply of raw materials were encountered by many pellet producers; even those with their own sawmills simply couldn't find enough raw material to satisfy rising demand. Production has risen enormously since then in all regions, so it is to be hoped that the problem will not arise again.

Any untreated wood waste can be used, from sawdust and wood chip, to larger off-cuts, whole trees and stripped bark. The feedstock must be from solid wood, never from composites such as chipboard. Different raw materials will require different levels of processing. Where freshly harvested entire trees are used, much more energy must be consumed during the drying process. This has economic, energy and environmental costs associated with it. Recycled wood can also be used, but not if it has been treated. Using treated wood as a source is limited by national regulations stipulating the amounts of particular compounds, heavy metals and additives that can be legally present in the pellets. The *Codes of Good Practice* (2000), developed by British BioGen for the DTI, state that pellets destined for large commercial use (mainly co-firing in power stations) can contain up to 15 per cent material other than biomass (MOB), while pellets destined for domestic use must be manufactured from virgin and clean wood only.

Pellet mills are able to accommodate varying species of wood, moisture content and form. Coniferous wood materials are most commonly used, although both softwoods (conifers) and hardwoods (such as oak and beech) can be used. The resultant pellet will possess the same density, calorific value and ash content for both soft and hardwoods. The rate at which hardwood and softwood release their energy, however, can vary (hardwood being much slower). Both the bark and xylem (the tissue which conducts water and minerals from the roots upwards) from trees can be used, although it is more common to use just the xylem because a high bark content produces more ash when burned. The lignin and resin in the wood act as natural binding agents, so normally no additives are required. Some form of starch or carbohydrate is commonly added to assist in binding the pellets at the pressing stage. Hardwoods (for example, willow chip) contain less lignin and thus additives such as starch may be needed to produce a high-quality pellet.

Raw material must be fairly dry but moisture content can vary from 10 per cent to 20 per cent by weight. Moist wood can be dried prior to pelletizing, while steam or water can be added to very dry wood to help with binding. The addition of vegetable oil can also be used to help lubrication. When a mixture of sources is used it may be beneficial to incorporate a blending stage into the production process. Before pelletizing, materials can be mixed (either crudely with a front-loader on a cement pad or very precisely using an automated metering system). This maintains a consistent input, reducing the need for adjustments to moisture and binding agent levels during manufacture, and guaranteeing a more consistent product.

The wood pellet manufacturing process

There are eight key stages in wood pellet manufacture:

- storage;
- cleaning;
- drying;
- grinding;
- pelletizing (pressing);
- cooling;
- screening;
- distribution.

Storage

The raw materials are delivered in batches but the pellet production process operates continuously, so a good storage system for the raw material is essential to keep the feedstock clean and dry.

Cleaning

This is particularly important for pellet plant that use scrap or recycled wood as a primary resource, and many pellet factories use screening and magnetic devices to clean up any plastic, large lumps of any kind or any metal that might be lurking in the delivered wood. Since it's impossible to remove other contaminants (such as heavy metals in treated wood), only clean wood can be recycled to manufacture pellets destined for domestic or small- to medium-scale use.

Drying

Although not all inputs require drying (for example, planer shavings), most do. Raw material with a moisture content of up to 20 per cent can be pelletized,

WOOD PELLET MANUFACTURE 89

however the optimal level is around 12 per cent or lower. Drum dryers fuelled either by gas or increasingly by waste wood are the most common type of equipment used. Drying consumes a large amount of energy and thus presents a challenge to the net energy value of wood pellets. The use of waste wood as a fuel is not only cheaper in the face of rising fossil fuel prices but also improves the green credentials of production, which is important when selling to markets such as Europe that have significant environmental regulations.

Grinding

Prior to being fed into the pellet mill, the raw materials must be given a homogeneous consistency. This can be achieved by various types of granulation machinery such as a hammer mill. Particles should be no larger than the diameter of the die-holes; any larger and the quality of the pellet is damaged. However, it is also possible to grind too finely, resulting in the loss of the fibrous quality of the material, which prevents the pellets from binding properly. The optimum consistency is something similar to breadcrumbs.

Pelletizing (or pressing)

Pellets are manufactured from the granulated material using a pelletizer (sometimes referred to as an extruder), which is usually a mill that presses the raw material into pellet form through ready-made metal dies. Large-scale purpose-built mills are the best at maintaining pellet quality, but there are plenty of examples in Europe and North America of mills intended for animal feed production that have been converted to manufacture wood pellets. The size and efficiency of mills vary but, roughly speaking, 100 horsepower provides the production capacity for 1 ton of pellets per hour.

Pressure is used to force the granulated wood through the holes in the pellet die, the size of the holes determining the diameter of the pellets produced. As the pressure and friction increases so does the temperature of the wood. The high temperature softens the lignin, allowing the fibres of the granulated wood to be compressed into cylindrical pellets. The size of the die is very important in creating the appropriate level of resistance and controlling the pressure and temperature conditions. If the holes are too large, the material simply slips through, whereas if the holes are too small the temperature becomes too high and the material is scorched.

It is during this stage that additives may be employed if necessary. By regulating the moisture level, binding agents and lubrication, ideal pellets of less than 10 per cent moisture and a density greater than $0.6 g/cm^3$ (0.02 pounds per cubic inch) can be produced from a variety of raw wood materials.

Cooling and screening

The pellets leave the pelletizing machine soft and hot and must be air-cooled using a cooler machine to allow the natural resins (or other binding agents) to set without caking. Once cooled, the pellets are passed over a vibrating screen to sieve out any fine particles and ensure the end product is clean and dust-free.

Figure 7.1 *Loading pellets into a delivery truck at Balcas, in Enniskillen, Northern Ireland*

Source: Balcas

These fine particles can then be reintroduced into the production system so no raw material is wasted.

Distribution

Your pellets are then ready for packaging and distribution, as seen in Figure 7.1.

Sweden's successful wood pellet supply industry

With Sweden's huge forest industry, it's hardly surprising that over 50 per cent of bioenergy in the country is used by industry. Most of this happens in plant close to the sawmills that produce the necessary waste or co-product. Sweden is far-sighted in this regard, having imposed VAT on energy as early as 1991 and introduced an Energy Tax Act in 1994. This tax was increased in 1996 and a special tax on nuclear power introduced. Investment grants were made available for biomass developments, so the scene was set for the emergence of a successful wood pellet manufacturing and supply industry.

Wood pellets were first produced, mainly from bark, in Sweden back in 1970 at Mora. Production costs were much higher than anticipated and there was at the time no market, that is, no combustion units to use the pellets. Mora pellet plant closed in 1986, although a competitor had appeared in 1984 when part of the Volvo group built a factory at Vårgårda; however, this only lasted three years. Another more successful manufacturer, the Municipality of Kil, started up in 1987. Here, the raw material was already dry and they have been producing 3000 tons of pellets or more a year ever since.

With the energy taxes introduced in the 1990s, pellet production capacity in Sweden grew to around a million tons a year by the turn of the century, with the newer pellet plants incorporating dryers and coolers into their pelletizing systems. With industry and the domestic sector increasingly convinced by pellet technology, demand for pellets was rising so fast that in 1998 Sweden even imported around 100,000 tons of wood pellets, mostly from Canada.

This partly reflects the inevitable chicken-or-egg scenario of many brand new industries; in such cases, it's very difficult to expand the supply without simultaneously expanding the market. Then it swings the other way and there's more demand than supply. The best scenario for everyone concerned – raw material suppliers, pellet manufacturers, boiler manufacturers and pellet end-users – is if this see-saw condition is avoided as far as possible.

In Scandinavia, foresters are among the best at getting wood out of the forests at low cost; in any case, sawdust waste had been an issue in the timber milling industry in the 1980s. Given this, wood pellets were considerably cheaper in Sweden by 2000, at €100 a ton (€21.27/MWh) than oil, at €45/MWh. Since 2005, pellet prices have remained less than half those of oil. To increase market penetration, one of Sweden's largest contemporary pellet manufacturers, Södra, offers clients heat rather than pellets or boilers. The customer pays for delivered heat (whether from a heat main or an automated pellet boiler on site), measured on a meter by the kilowatt-hour. Pellets are easy to buy in Sweden either in bulk, delivered by a blower lorry, or in big bags, while for stoves, they can be picked up in 10–35kg (22–77 pound) bags from petrol stations across the country.

By 2008, Sweden had developed a much larger domestic market and there were over 94 pellet production plants. Stove sales rose from fewer than 1000 in 2000 to more than 14,000 in 2006. The purchase of pellet burners under 25kW rose from fewer than 20,000 to well over 100,000 in the same period. Simultaneously, the sale of pellet systems between 24kW and 500kW grew from around 600 to over 3500. To meet all this demand, Sweden produced over 1.2 million tons of wood pellets in 2006 and developed efficient pellet distribution networks to sell the product through. The future looks bright, although there are perhaps two main risks: the development and improvement of alternative heating technology (for example, ground-source heat pumps); and possible future competition for the raw material, which is a finite resource, even in Sweden.

References

British BioGen (2000) *The British BioGen Code of Good Practice for Biofuel Pellets and Pellet Burning Appliances <25kW*, Version 1, British BioGen, London, www.woodfuelwales.org.uk/biomass/Technology/cogpp.pdf (accessed 23 November 2009)

The Log Pile website, www.nef.org.uk/logpile/pellets/production.htm (accessed 23 November 2009)

Austria, ProPellets website, www.propellets.at

8
Marketing Wood Pellet Technology

The market for wood pellets

Estimates vary, but they suggest that as many as 8 million tons of wood pellets were produced in Europe during 2008, while in North America a further 2 million tons were manufactured. At the cheapest sale price of €100 a ton, wood pellets can be conservatively estimated as a €1 billion global market. Although wood pellet stoves and small-scale boiler technologies are used in all markets, it is larger-scale use, mainly in power stations, that has dominated the industry's growth in Scandinavia and North America. But as the market matures, even in these countries, the domestic sector is also establishing itself as significant.

In Europe, where there are over 300 pellet manufacturing plants, Scandinavia, Austria and Germany have led pellet market development, with other countries such as Italy, France and the UK following their lead. The company Balcas now produces over 155,000 tons of wood pellets a year between its base in Northern Ireland and a new plant in Scotland.

As well as the progressive approach of European Commission technology programmes (such as Intelligent Energy for Europe), the main drivers for pellet market development have been environmental concerns, rising fossil fuel costs and compatibility with a large associated forest industry. The order of priority for these drivers has varied from region to region and it's true to say that, although perhaps predictable, the increasing price of oil and gas has added recent momentum to market growth everywhere. In the UK, where we have relatively little forest – although wood pellet technology development clearly offers new markets for foresters and sawmill owners– the forestry sector lobby has been less powerful a factor. Nevertheless, the UK Forestry Commission has been facilitating capital investment subsidies for both wood fuel production and combustion units since the year 2000, which has helped to kick-start the market here.

With heat accounting for around half the energy used in Europe, perhaps it's not surprising that one Austrian company – the Andritz Group – claims a 50 per cent share of the world market for wood pellet production equipment. As a whole, the Andritz Group operates in several related markets for equipment and services: pulp and paper, biofuel, steel, animal feed and hydro power. According to a study by this group, in 2008 alone wood pellets replaced around 6.3 million tons of coal. The USA uses more than 900 million tons of coal a year to help generate electricity, so there is plenty of scope yet for market

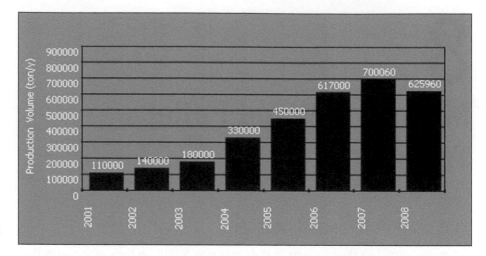

Figure 8.1 *Wood pellet production in Austria*

Source: Pellets@Las Project (Rakos, 2008)

growth here. Canada, too, has a large forest-based industry and pellets are being increasingly used at a large scale. In Europe, many of the eastern countries and new European Union (EU) member states are keen to develop wood pellets since they have both forest industries and lots of very inefficient fossil fuel-fired power stations and heating systems for the residential sector.

The global pellet market is perceived as displaying exponential growth and has attracted significant investment over recent years, from both the industry's and the end-user's perspectives. With the EU seeing wood pellets as one of the keys to achieving its targets for renewable energy and carbon emission reductions and the Obama administration in the USA desperately looking for quick ways to reduce power station emissions, the market is likely to keep expanding rapidly for at least the next decade. The limits to market growth will be largely determined by:

- the availability of raw material and competition for this resource;
- development of alternative clean heating sources;
- continued political enthusiasm for the technology and the provision of fiscal policy instruments.

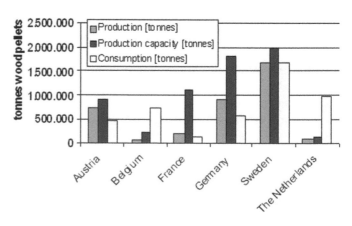

Figure 8.2 *Comparison of production, production capacity and consumption in Europe 2007*

Source: Pellets@Las Project

Marketing the technology

Marketing wood pellet technology has varied in style and emphasis from country to country and company to company, but the example of Austria stands out. Here, there has been a successful combination of technology promotion alongside R&D and commercial development.

Without partnership between sectors, including the EU and national and regional government funding for support agencies, it would have been difficult to market pellets so effectively. The emphasis on fuel quality standards, developing working distribution networks and getting information to potential end-users and those who make decisions about building heat were all essential to the creation of a successful pellet market in Austria. All required cooperation between these main partners.

Establishing a new market for pellets requires three main elements: pellet fuel, pellet combustion equipment manufacturers (the means to use the fuel) and pellet end-users (demand for the fuel). Failing to address the supply and demand sides simultaneously is a recipe for at least temporary havoc in the market. It's not enough to produce wood pellets and wait for people to wake up to the idea of burning them. People need to be convinced at every level, not just by manufacturers' sales and marketing teams, but also by government and energy advice agencies (for example, the highly proactive and successful Upper Austrian Energy Agency – O. Ö. Energiesparverband or ESV). Making pellet combustion units without pellets to burn would be a non-starter; and no household or large building owner is going to invest unless they have access to both pellet burning equipment and a pellet that is consistent in price as well as quality.

In Austria, the supply side has received significant support for capital investment in plant and also for R&D. Taking matters further, plumbers and heating installers were targeted for awareness raising and training with support from energy agencies. ESV started and still organizes the annual European Pellet Conference alongside the World Sustainable Energy Day events in Wels, Upper Austria. In this way, it has helped establish Austria's strong position within the world pellet market, which helps with exports, and also spreads its expertise and good practice.

The rapid growth experienced in the Austrian pellet market slowed during 2007, although it recovered well until pellet prices rose again as an indirect effect of the financial crisis that started in 2008. A combination of bad press in 2006, possibly initiated by competitive interests in the oil and gas sectors, and higher pellet prices at that time resulted in a lower than expected take-up of new pellet combustion systems. As a result, 2007 saw a massive oversupply of pellets at the same time as some large-scale investments in new boiler production factories were just being completed. Instead of the market for pellets being overheated, it began to hiccup, with a lack of demand for equipment and slower growth in pellet sales. To make matters worse, the 2006–07 winter was a mild one, so far fewer pellets than projected were eventually sold in Austria.

Later in the year, pellet prices stabilized well below those of oil and ProPellets Austria organized a series of successful PR events and positive media stories. Press coverage reverted to highly positive and the demand for both pellets and pellet combustion units is now rising again. This illustrates what is perhaps the single most important factor for the success of marketing pellets and pellet technology in Austria – the fact that agencies such as ProPellets and ESV are closely monitoring the state of play in the market, working all the time to keep it as balanced and buoyant as possible. Unlike other stakeholders, they maintain an objective overview and integrative role that can fine-tune the market to some extent, shifting emphasis from pellet production capacity and pellet quality to equipment supply or consumer demand according to the situation on the ground. However, pulling the strings of the various market sectors isn't quite as easy as it sounds; for instance, it usually takes anything from one to two years to get a European-funded R&D or capital grant project off the ground. In this light, support from local and regional authorities, which can sometimes be more rapidly forthcoming, also plays a vital part in the process of managing markets. In 2008, the main support for pellet heating was coming from the provincial level in the form of €2000–3000 grants from energy and climate funds. At the federal level, an additional subsidy was introduced for up to €800. With pellets now half the price of oil in Austria, consumer interest is growing once again.

The ÖkoFen example – excellent design plus vertical integration

Delighting the customer with efficient and reliable wood pellet boilers

The bottom line for successful pellet technology marketing depends on the customers – whether residential, commercial or public – having high levels of confidence in both the combustion technology itself and the consistent supply of pellet fuel. Several Austrian wood pellet boiler manufacturers have made huge efforts to address the former, focusing on designing and manufacturing smart-looking, efficiently operating and intelligent boilers that can meet market requirements. ÖkoFen, for instance, supplies products from eco mini boilers right up to 224kW energy houses that can be delivered to site within six weeks of being ordered and up and running in just a day.

Founded in 1989, ÖkoFen now has around 260 employees, including subsidiary companies. Some 80 per cent of its manufactured product is exported (mainly to neighbouring Germany). It produces an integrated range of pellet boilers, but the key to its marketing success is probably its emphasis on quality (example of boiler is shown in Figure 8.3).

From the outside, it certainly looks as if its strategy has been based on horizontal integration as the key to ensuring product and system quality. It has either set up or taken over companies to produce the entire spectrum of complementary products such as solar collectors, buffers and combi tanks and pellet storage systems. It also has its own software for the microprocessors in its units. In this way, all the component parts are known to be totally compatible.

Ensuring quality of fuel used in boilers

Of course, to be really sure its boilers operate efficiently, ÖkoFen is quite strict with its recommendations for wood pellet fuel quality. It has been involved in the process of establishing pellet quality standards in Austria. Uniform shape and size for pellets is important to ensure optimum heat output and consistency in the flows required for the fuel from factory to end-user's store and store to combustion

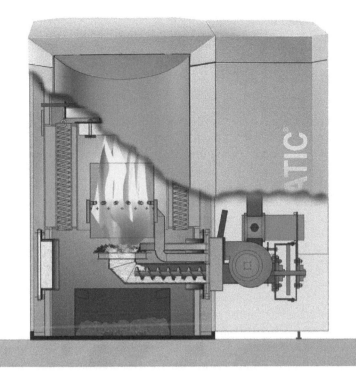

Figure 8.3 *ÖkoFen PEL-LEMATIC boiler*

Source: ÖkoFen

unit. ÖkoFen recommends wood pellet sizes between 5mm (0.20 inches) and 6mm (0.24 inches) in diameter and 10–25mm (0.39–0.98 inches) long, with a minimum density of 650kg/m³ (40 pounds per cubic foot). Pellets' length will have a negative impact on their bulk density; the consequence of this is that for every turn of an auger there will be fewer pellets the longer they are. If they are longer than 25mm (0.98 inches), there is a chance of the supply ducts becoming blocked as a long pellet somersaults its way along the delivery tubes to the burner.

Projecting an environmentally friendly and luxurious image

For marketing purposes, ÖkoFen use a happy and relaxed picture of a young child lying very happily on top of (and partly submerged in) a mountain of wood pellets. The child is a family member of ÖkoFen's founder and MD, Herbert Ortner, which perhaps adds emphasis to this clear illustration of how people-friendly and tactile this fuel is. It certainly beats seeing a child lying in a tank of oil.

Figure 8.4 *Child in pellet store*

Source: ÖkoFen

This very strong image of wood pellet fuel was selected to overcome perhaps the biggest single factor effecting market growth – the confidence of and positive perception by end-users and potential end-users. It also fits in well with the company's image of selling stylish boilers for a luxurious lifestyle. This kind of image projection has been much more common in Austria than in the Nordic countries or North America where it has more traditionally had a more rustic or industrial image.

Country example – Austria

The first wood pellet systems in Austria were introduced in 1997. Since then, the industry has seen remarkably rapid growth. Between 2003 and 2007, pellet production quadrupled with this upward trend continuing to the present time. Around 30 pellet producing companies are now currently active in Austria, with a total production capacity of over 800,000 tons per annum.

Woody biomass (wood pellets and wood chip) now accounts for a 12 per cent share of Austria's primary energy consumption. Around 60 per cent of this is consumed just in the domestic sector. In Austria, bioenergy has been traditionally based in the residential heating sector, which, because of the significant forest cover, generally has close local ties with the wood processing industry. There are more than 16,000 wood pellet central heating systems operating in Upper Austria and no country in Europe has a comparable level of market penetration in pellet boilers. However, more recently commercial programmes have been introduced to promote the adoption of biomass heating systems by businesses. The Austrian government hopes that these subsidies will encourage businesses to take up the more expensive pellet boilers. If this works, it has the potential to generate considerable additional domestic market growth. The government is reported to consider the EU's renewables and carbon emission reduction targets highly ambitious; nevertheless, it continues to pull its weight at home. With energy agency and government support, the Austrian pellet boiler manufacturers have now set their sights on the world of exports beyond their immediate neighbours such as Germany and Italy.

The most significant force driving the success of the wood pellet industry in Austria is the government's strong political commitment to environmental concerns. The ratification of the Eco-Electricity Act in 2003 played an instrumental role in the growth of the bioenergy industry in Austria. This act aimed to increase the share of renewable energy sources to 78.1 per cent of Austria's total energy consumption by 2010 and, according to a study by the Austrian Energy Agency, the government hopes to create additional demand for bioenergy amounting to at least 2 million m^3 of wood fuel per year.

An abundance of sawmills and other feedstock-producing industries provided Austria with high levels of forestry co-product (which would otherwise be waste wood) that were necessary to meet increasing demand with minimal need for importing raw materials or plant. Over the last five years, this high production capacity has also enabled Austrian companies to take advantage of the large export market in Europe. This is likely to extend soon to North America where combustion technology has been slower to develop and improve. In 2005 around 40 per cent of Austria's total pellet production was taken up by exports with 150,000 tons exported to Italy and 45,000 tons to Germany. Two years later, following a mild winter, Austria experienced an unexpected oversupply of pellets.

Production and consumption of wood pellets at both the residential and commercial scale has been publicly promoted through initiatives such as the

Upper Austrian Wood Pellets Programme, managed by ESV and supported financially by tax incentives and subsidies. The introduction of quality standards and regulations such as the ONORM M 7135 has played a significant role. At the very least it has helped to give customers (at the individual, commercial and international level) and prospective customers that vital confidence in the availability and quality of the product.

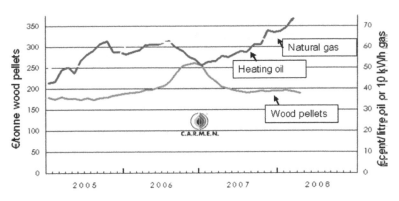

Figure 8.5 *German wood pellet and oil prices compared*

*Source:*CARMEN

There has also been a great deal of investment in technology development leading to a wide range of market-ready combustion units being available locally. R&D has increased the efficiency of bioenergy systems dramatically from an average of 50 per cent in 1980 to over 90 per cent now. At the same time, emissions have decreased to values below 100mg per normal cubic metre (Nm^3), making wood pellets a more economically and environmentally viable option. Austrian companies have also appealed to consumers by investing in the aesthetic design and user-friendliness of their boilers, with some even providing a wide selection of colours. In addressing the need for renewable energy sources, Austria has taken a multifaceted approach, setting itself up as one of the most successful wood pellet producers in Europe.

Economics and finance

The economics of wood pellet markets has generated significant interest over the last few years, making finance more readily available than ever. Within the EU, biomass in general, and wood pellets in particular, have helped achieve major inroads towards offsetting fossil fuel use, even when compared to other renewables. The data comparing pellet and oil prices (see Figures 8.5, 8.6 and 8.7) indicate an increasingly competitive scenario for biomass.

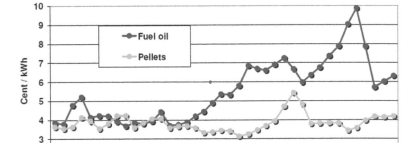

Figure 8.6 *Austrian wood pellet and oil prices compared*

Source: ProPellets Austria (Rakos, 2008)

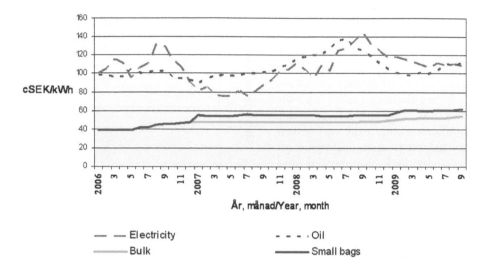

Figure 8.7 *Swedish wood pellet, oil and electricity prices*

Source: Pellets@Las Jonas Höglund (2008)

The importance of price

The price of wood pellets for heating now compares more favourably to those of oil and even gas. Since price is assumed to be the most important factor ultimately determining which way consumers will spend their money, things look pretty good at present for the wood pellet industry and this has been creating significant excitement in the investment circles of both Europe and North America for several years now. However, this has not always been the case.

Unlike many other renewables – such as wind, hydro and solar power – someone has to grow biomass and then go out to harvest the fuel before transporting it for pre-user processing. In the case of pellet fuel, the wood is delivered in batches to a pelletizing plant. This adds to the jobs created by the industry as a whole, but it also adds an additional cost element over other clean natural energy sources. Neither is this raw material's cost fixed. It's not difficult to imagine a scenario where other industries compete for the wood co-product essential for making pellet fuel. Sawdust prices increased from €4 to €15/m³ between 2004 and the second half of 2006. This fact was certainly a contributing factor to the bad press experienced by the Austrian pellet industry in 2006, temporarily damaging the cosy image of wood pellets. Stove and boiler sales dipped while capacity for production was at its highest and still growing based on unexpectedly high sales over the very cold winter of 2005–06.

At the time of writing, the price of pellets is highly competitive and it's hard for even sensationalist press to criticize the industry in any way. However, there were reports in July 2009 that despite a slight decline in the pulping industry the pellet industry is having increasingly to compete with both wood-based panel manufacturers and pulp mills for wood residues and logs. The price of small round wood thinnings has gone up significantly in both Germany and Sweden, mainly as a result of wood's increasing use for energy generation.

The price of fuel is not the only factor influencing purchase

As well as the ongoing expenditure on fuel, as with the purchase of a car, buyers have to weigh up the differences between options in terms of the capital costs involved in meeting their heating needs and objectives. Wood pellet systems are generally a lot more expensive than oil or gas systems of comparable output from this initial investment perspective.

However, the capital costs may be less relevant to a potential heating system buyer who is more concerned with style, reliability or simply being green. Austrian stoves and boilers, despite generally costing more, are still doing very well at home and abroad. This is largely because of their reputation for reliability – being perceived, perhaps, as the producers of the best-engineered pellet combustion equipment; but it is also related to style, with the top Austrian manufacturers sometimes referred to as producing the Rolls-Royces of pellet boilers.

Often, the upfront capital costs involve more than just the pellet boiler. Where a property is transferring from another fuel to pellets, it will need to consider the costs of fuel storage and fuel conveyance systems too. These add considerably to the overall investment costs. This has undoubtedly been a drag factor on market penetration for pellet systems and pellet fuel.

The case for subsidies

In Austria (see country example above) and in Germany, there have been high enough levels of subsidies to achieve substantial market growth. Norway has had problems in developing the industry because of much lower percentage investment subsidies. In Denmark, which has a relatively well-developed pellet market, good subsidies are available for pellet heating systems, but only to properties not connected to a district heating main.

Typical cost breakdowns for wood pellet systems

The first matter to clarify about cost is what we mean by 'wood pellet system' in this context. Few clients just buy a pellet boiler alone from a supplier. If one does, there is much less likelihood of the system working well and no boiler manufacturing company would guarantee the whole system anyway. It is always advisable to stick as far as possible to one single supplier and contractor. The detailed description of a wood pellet system (see Chapter 4) includes a host of devices that are not needed for oil or gas heating systems. Given this, heating systems that are being converted from conventional fuels such as oil or gas will require a one-off investment in the full system. The wood pellet boiler system requires many more elements than most conventional boilers:

- pellet store;
- pellet conveyance system;
- sophisticated controls;
- accumulator (perhaps);
- fuel system;
- heating circuits.

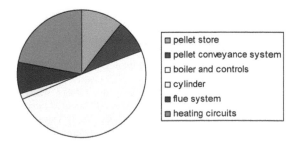

Prices vary enormously between wood pellet boiler system types and makes, but one rule of thumb is to expect a pellet boiler system to cost between £1000 and £1500 for every kilowatt of installed capacity.

Figure 8.8 *Cost breakdown of a typical wood pellet boiler system*

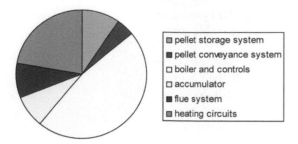

Figure 8.9 *Cost breakdown of a small school installation in Wales*

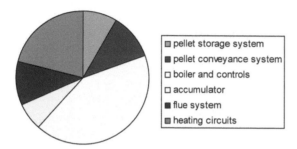

Figure 8.10 *Cost breakdown of a TP30 boiler system installation*

Source: EcoEnergyDepot, UK

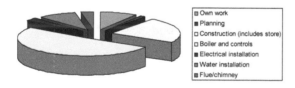

Figure 8.11 *Cost breakdown according to Austrian experts*

Source: Regionalenergie Steiermark

With this formula, you could expect a 10kW boiler system to cost from around £10,000. Judging from Figure 8.8, a £10,000 boiler would make up round 50 per cent of the costs of a £20,000 system, with the second half, in order of descending cost, largely comprising the heating system, pellet store and flue system. Where there's an accumulator tank in the design, the pellet conveyance system is usually the least costly main element; where there isn't a tank then a cylinder would be the cheapest main element.

It is easy to see that the boiler and controls form the major part of this typical domestic wood pellet boiler installation, followed by the wet radiator heating circuits, if not already in place, and the pellet store. Both the flue and conveyor system are significant expenses, but relatively cheaper.

Figure 8.9 shows a two-boiler installation providing space heating and hot water for a small rural school in Powys, Mid Wales (see the school case study in Chapter 9 for more details). The cost breakdown is pretty similar, particularly the split between costs of boiler and controls and the heating circuits. The use of two accumulator tanks in the system adds an extra and significant cost element.

Obviously, the breakdown and exact costs will vary between supplier and boiler system type, but, again, based on actual costings for a 20–30kW domestic boiler system (see Figure 8.10), we see a very similar pattern. However, the vacuum feed conveyance system cost element is noticeably proportionally a little larger than the typical system figures in the actual school system cost breakdown.

Taking a different perspective and including elements such as one's own labour, but not the heating distribution system costs, a study by Austrian consultants Regionalenergie Steiermark (see Figure 8.11) ranks the system elements as percentages of the overall system price.

References

Höglund, J. (2008) 'Current Pellet Market Developments in Sweden', paper presented at European Pellet Conference, Wels

IEA (2007) *Global Wood Pellets Markets and Industry: Policy Drivers, Market Status and Raw Material Potential Bioenergy Task 40*

Rakos, C. (2008) 'Development of the Austrian Pellet Market 2007', paper presented at European Pellet Conference, Wels

Regional Energie website, www.regionalenergie.at

9
Case Studies

Case study 1 – 5.5kW pellet stove

Figure 9.1 shows a 5.5kW Extraflame Baby Fiamma wood pellet stove, installed in a large house near Fort William, Scotland, which provides a convenient heat source to augment the existing central heating system. The householders wanted to install an automated system with a relatively low running cost and low maintenance requirements. They also wanted an aesthetically pleasing stove with a real flame. Wood pellet fuel is purchased locally in bags that weigh 10kg (22 pounds) and cost £1.80. This corresponds to heat at 3.75 pence per kWh.

Features of the system include:

- induced draught flue fan for smokeless operation;
- automatic ignition;
- very low maintenance requirements;
- remote control operation for temperature and output settings;

Figure 9.1 *Extraflame's Baby Fiamma 5.5kW wood pellet stove*

Source: Highland Wood Energy Ltd, 2009

- heating time programmer for individual weekdays with holiday function;
- room thermostat, temperature adjustable by remote control;
- stove output adjustable by remote control;
- programmable day and night temperature settings.

Case study 2 – 14.8kW wood pellet stove with back boiler

An Italian-made Extraflame Ecologica Idro model wood pellet stove was installed in a family home in Llanidloes, Mid Wales, in December 2004 (see Figure 9.2). The intention was to provide all the space and water heating needed for a three-bedroom Victorian house. Previously, the occupiers had struggled to keep the place warm in winter, using wood logs on a Rayburn stove. The Rayburn cost around £400 a year to fuel with logs, but that was five years ago.

Since installation, the new wood pellet stove distributes its heat with a back boiler connected to a system of nine radiators running throughout the house. The house is much warmer than it used to be, which the occupier puts down to the pellet heating system with its much higher combustion efficiency. However, the user believes that about half of the heat is given off directly into the air, and would prefer it if more went into the radiators. He has independently arranged for heat to circulate around through air vents in the house.

The stove has a seven-day programmer with digital display but offers only two time period settings, which means that it cannot be programmed for varying periods between weekdays and weekends. You can also program it if you are going away for up to 15 days; it will not then fire up until you return, or the day before, depending on user programming. Pellets are auger-fed and the stove has automatic ignition.

The Extraflame pellet stove presently consumes some 2.5 tons of pellets annually. These are sourced locally in bags, two or three times a year, at about £250 a ton, including delivery. The pellet stove and installation cost around £6000 including the flue, which has a 125mm-diameter twin-wall flexible flue liner. The stove itself cost £2750 (plus VAT). The house also has a solar thermal input from a single flat-plate collector installed by a local company, to augment the domestic hot water needs.

Figure 9.2 *Extraflame's Ecologica Idro 14.8kW wood pellet stove with back boiler*

Source: Andy Warren

Case study 3 – pellet central heating boiler with solar-assisted domestic hot water in low-energy new-build house

Figure 9.3 depicts a super-insulated new-build detached house in temperate climatic conditions located in rural Wales. It is a wooden-framed building with three bedrooms and a larger open-plan ground floor with living spaces and kitchen. Domestic hot water is assisted by a small flat-plate solar collector positioned on a south-facing roof (see Figure 9.4).

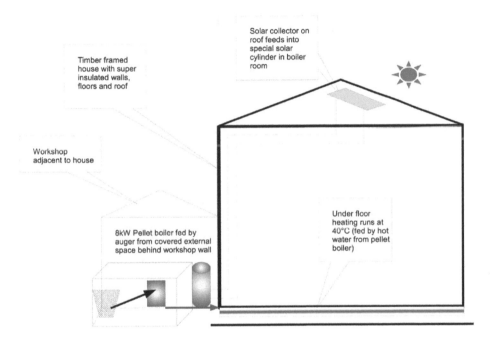

Solar collector on roof feeds into special solar cylinder in boiler room

Timber framed house with super insulated walls, floors and roof

Workshop adjacent to house

8kW Pellet boiler fed by auger from covered external space behind workshop wall

Under floor heating runs at 40°C (fed by hot water from pellet boiler)

Figure 9.3 *Sketch of pellet central heating boiler with solar-assisted domestic hot water in low-energy new-build house*

Located in the centre of a village, the house is not particularly exposed to local weather conditions. The windows and doors are triple-glazed. The roof and floors are insulated to a significantly greater degree than the recommended minimum (in Part L of the Building Codes; see Chapter 5) for the UK. Exterior walls contain 145mm (5.7 inches) solid foam insulation. There are also plans to install a heat recovery system, which would make the property even more energy-efficient.

The aim of this boiler installation was to minimize pellet fuel consumption by maximizing both thermal and combustion efficiencies. An 8.2kW boiler was installed in a garage-sized outbuilding adjacent to the

Figure 9.4 *Photograph of low-energy house with pellet boiler and solar thermal array*

Figure 9.5 *Photograph of mixing system and pipes for under-floor heat distribution*

house. The same outbuilding houses a solar hot water cylinder and pump stations. A twin-wall stainless steel flue was installed in the boiler room. Heat is distributed via a multi-zoned underfloor coiled system (see Figure 9.5) that operates at only 40°C (104°F) for the ground floor; upstairs there are steel wall-mounted radiators, each with independent thermostatic valves. There are also 4m² (43ft²) of flat-plate solar collectors, selected for compatibility with the pellet boiler and automatic control system.

The pellet store is a flexible tank located directly behind the boiler, but on the other side of the external wall in a separate covered space that can be filled from an open yard belonging to the adjacent property. Wood pellets are fed by screw auger through the external wall into the boiler's combustion area. To date, the pellets have been delivered in bags and the flexible tank has been filled by hand, partly to check the quality of the wood pellets during the first year of operation. Bulk pumped tanker deliveries are, however, possible.

According to the property owner, the 8.2kW pellet boiler could heat a significantly larger space, given the same very high levels of insulation and similar climatic conditions and building exposure levels. The house is warm all year and the boiler only fires up a few times a day, ensuring low levels of pellet fuel consumption. After the first winter, some of the initial 3 tons of pellet fuel still remained unused in the store. It is quite possible that a smaller boiler could effectively heat this thermally efficient house.

Case study 4 – pellet central heating boiler with solar-assisted domestic hot water in standard house

Figure 9.6 depicts an average-sized (three-bedroom) detached house in temperate climatic conditions located in rural Wales. Domestic hot water is assisted by a small solar thermal collector positioned on a south-facing roof (see Figure 9.7). There are two adult occupants with occasional visits by relatives and friends.

Although this is an older property, insulation has been fitted in the ceilings, roof and floors. The roof has 300mm (11 inches) of mineral fibre insulation (reduced to 75mm (3 inches) around the sloping ceilings). Underneath the solid floors there is a 30mm (1-inch) layer of foam insulation. The property is double-glazed (K-glass type). There is a large pre-existing chimney in the house, through which some air is lost. The previous heating system comprised electric storage heaters augmented by a wood stove in the main living room. In the depths of winter, the occupants were rarely as warm and comfortable as they wanted to be.

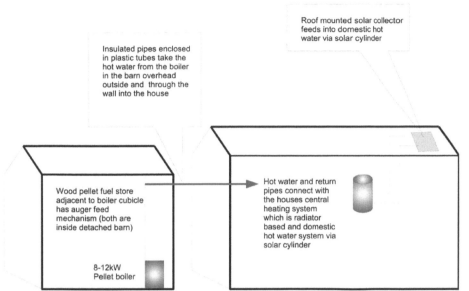

Roof mounted solar collector feeds into domestic hot water via solar cylinder

Insulated pipes enclosed in plastic tubes take the hot water from the boiler in the barn overhead outside and through the wall into the house

Wood pellet fuel store adjacent to boiler cubicle has auger feed mechanism (both are inside detached barn)

Hot water and return pipes connect with the houses central heating system which is radiator based and domestic hot water system via solar cylinder

8-12kW Pellet boiler

Figure 9.6 *Sketch of pellet central heating boiler with solar-assisted domestic hot water in standard house*

Figure 9.7 *Photograph of standard house with pellet boiler and solar array*

The boiler installed is rated at 15kW and has been sited next to a flexible tank pellet store inside a barn just a few metres from the house. A screw auger feeds pellets from the store into the combustion area of the boiler and the hot water produced is delivered by well-insulated overhead copper pipes from the boiler into the house where it is fed into a heat distribution system that uses standard wall-mounted steel radiators with independent thermostatic valves. No accumulator was installed. A twin-wall stainless steel flue was installed in the

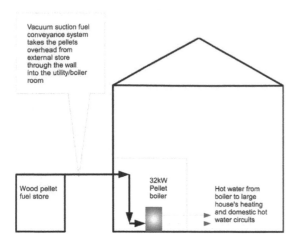

Figure 9.8 *Sketch of large domestic pellet central heating boiler*

Figure 9.9 *Photograph of country house with wood pellet boiler*

boiler room. The occupants are now warmer and more comfortable in the winter months and there is a high level of customer satisfaction.

The boiler has a summer setting for domestic hot water that fires up twice a day to suit the occupants' lifestyle. If the weather is particularly sunny and domestic hot water needs are met by the evacuated-tube solar thermal array, then the boiler will not fire. Domestic hot water is heated in a twin-coil solar cylinder.

Wood pellets were originally sourced locally but were not of the quality required. Better-quality pellets are now bought from a new pellet manufacturer located about 130km (80 miles) away. Deliveries are in bulk by tanker, approximately every nine months. The property consumes between 5 and 6 tons of pellets annually.

Case study 5 – large domestic pellet central heating boiler

This case study focuses on a system at a large Victorian property that also functions as a country guest house in a tourist region (see Figure 9.8). It has four double bedrooms, high ceilings and large rooms throughout. Based in the temperate climatic conditions of rural Wales, the house is protected from potential exposure to winter weather by nearby trees and a hill (see Figure 9.9). The southern aspect, also shaded by trees, was not conducive to the installation of a solar collector, so all domestic hot water is provided, along with the space heating, by the wood pellet boiler. There are two adult occupants with frequent additional occupancy by client-guests.

The building's exterior comprises 50mm-thick (2 inches) stone external walls without insulation. The roof space is insulated with 150mm (6 inches) of mineral fibre and double-glazing was installed at the same time as the pellet boiler. The heat distribution system is based on wall-mounted steel radiators with independent radiator thermostatic valves.

A 32kW wood pellet boiler was specified to heat the entire house, including guest rooms. The boiler's ability to modulate means that it is effective and efficient whether the house is part-occupied or fully occupied. The boiler is located in a utility room along with washing machines and other household appliances close to the store, and a new stainless steel double-skin flue was installed, leaving the house through the utility room roof (see Figure 9.10). There is no accumulator tank in the system.

The pellet store is located in a stone shed, about 2m (6 feet) apart from the main house. Pellets are fed via an overhead vacuum suction tube directly into

Figure 9.10 *Photograph showing camouflaged overhead vacuum pellet feed tube going from stone pellet storage room on the right into the boiler/utility room roof on the left*

the boiler. Fuel is delivered in bulk by trucks that blow the pellets directly into a flexible tank. Occasional blockages have been experienced with pellet dust accumulating at the bottom of the flexible tank. Furthermore, the large size of the first tanker to deliver pellets did cause some damage to the driveway. Smaller trucks now deliver their pellets in bulk, usually 3 tons at a time, and since switching to this new pellet supplier, pellet quality has not been an issue.

This property was previously heated by LPG, which was delivered by tanker. Since the introduction of the wood pellet boiler, heating costs have been significantly reduced and the comfort levels have improved considerably, particularly during extreme winter weather conditions. The guest house mentions its wood pellet heating in its promotional literature to target environmentally friendly tourist clients.

Case study 6 – school wood pellet central heating boiler

This case study looks at a school wood pellet system installed in Powys, Wales (see Figure 9.11). Work began in October 2007 and was completed in July 2008. The system was based on two pellet boilers, both rated at 56kW, operating in tandem with a potential output range of 17–112kW (see Figure 9.12). The school heat distribution system is a variable-temperature underfloor circuit operating at temperatures up to a maximum of 50°C (122°F). For the kitchens and IT suite, there is a constant-temperature air handling system, which, along with the primary hot water circuits, operates at 80°C (176°F).

The control system has remote web-accessible TAC building management system (BMS) (county standard). An accumulator tank (buffer vessel) is integrated into the system primarily for winter use; the council is considering

Figure 9.11 *Photograph of new-build school heated by wood pellets in Wales*

Figure 9.12 *Photograph of school boiler room with two 56kW boilers installed*

bypassing this for summer primary low loads. The boilers have their own feed pumps to serve the buffer vessel, ensuring that the boilers operate independently and at the best efficiency with pump overrun for heat dissipation. The primary (hot water) circuit is augmented by a demonstration size (4.6m², or 49.5ft²)) solar collector area. There are no back-up or stand-by fossil fuel boilers. The wood pellet silo (see Figure 9.13) is housed in a separate building with pellets delivered 10m (32ft) without any problems via a vacuum suction system.

Insulation levels were to UK Building Codes Part L, 2006 standard. The floor area is 832m² (8955ft²) and the ventilation factor around 48w/m². Overall the heat loss is 40kW, plus primary load and air handling heater battery load. The buffer tank holds 3000 litres (792 gallons) (based on a rule of thumb that

Figure 9.13 *Photograph of school wood pellet silo*

20–40 litres (5–10 gallons) are needed per kilowatt of system load). The boilers were sized for peak loads and also to provide back-up in case one unit breaks down.

Problems experienced to date include:

- over-compaction of the fuel following poor pellet delivery procedure;
- seizing-up of the feed auger (a consequence of the dusting of poorly formed pellets);
- unclear arrangements for day-to-day operation and maintenance (e.g. ash removal).

There were initially a few minor problems with the boilers and pellet deliveries from suppliers, but it nevertheless worked well through the first winter, with just a couple of minor breakdowns that were quickly rectified. The school staff are happy with the installation.

School heating is the responsibility of local authorities in the UK and in the county of Powys some seven schools have now been converted from oil boilers to wood pellet systems. Powys County Council has found that pellet boilers, compared with wood chip boilers, require less intervention by virtue of their automatic operation, clean fuel and simpler fuel delivery and combustion systems. Occasional ash removal and twice-yearly servicing are the main requirements. Servicing costs are similar or slightly higher than fossil fuel-fired boiler plant, with negligible repair costs to date.

Pellet biomass boilers installed in Powys during 2008 and 2009 initially suffered from poor fuel quality and delivery problems. Two Welsh pellet manufacturers failed during 2008, but a new supplier from North Wales now provides good-quality pellets and the operation of the pellet boilers has improved dramatically. More school pellet boilers are being commissioned in the region. These are facilitated by 50 per cent grant funding towards the capital costs of installing biomass boiler plant for the next two years.

The relative strength of the euro over the pound in 2009 means that the capital cost of the European manufactured boiler plant and equipment has risen by about 15 per cent. However, the Renewable Heat Initiative comes into force in 2011 and is projected to pay between 2 pence and 3 pence/kWh of heat produced, making the economics of wood pellets better still. Furthermore, the Carbon Reduction Commitment will charge in the region of £12 per ton of carbon dioxide emitted. Since biomass installations are said to be carbon-neutral, the new school heating systems will avoid this cost. Biomass fuels are also zero rated under EU emissions trading schemes.

Case study 7 – wood pellet CHP

Today, Sweden's largest user of wood pellets is Hässelby CHP plant (see Figure 9.14), which uses between 250,000 tons and 300,000 tons a year. One of the earliest plants anywhere in the world to convert from fossil fuel to biomass, the project was critical for future large investments in pellet manufacture in northern Sweden. Its location in the northwestern part of Stockholm on the shore of Lake Mälaren facilitates large-scale delivery by boat from Sweden's northern forest products industry. On arrival, pellets are easily unloaded into a massive storage area. Now owned by Fortum Power and Heat AB, the plant provides heat (500GWh for over 45,000 homes) and power (260GW of electricity) for the northwestern suburbs of Stockholm.

Figure 9.14 *Photograph of Hässelby CHP plant*

Source: Eddie Granlund, Fortum

Although it is Sweden's first ever CHP installation, Hässelby was initially designed in 1959 to be fuelled by oil. Following the energy crisis of the 1970s, however, it was converted for coal burning in 1983. Ten years later, when environmental permits for coal burning came to an end, the plant's owner, then the City of Stockholm, decided to switch to wood pellet fuel, making it the first biomass-fired, cogeneration power plant in Stockholm. The shift to wood pellets was achieved in several stages, not least because supplies of pellets were limited in the early years. By the early 2000s, Hässelby was running all of its three 100MW boilers on 99 per cent renewable resources, almost entirely wood pellets. The carbon dioxide emissions saved annually by the project are calculated to be in the region of 430,000 tons (Niininen et al, 2009).

Stored in large indoor bunkers, the pellets are delivered by a large screw auger device from the bunker to a conveyor belt, which carries them into a grinding mill. Here, the pellets are pulverized, much like coal, to prepare them for combustion in the large boilers, which are also linked to electricity generating turbines.

Large-scale heating or CHP plants require district heating networks (DHNs) where a central plant provides heat to a variety of consumers around the local area. Insulated pipes are used to transfer the heated water underground (usually 60–80cm (23–31 inches) below the surface) to all its customers. Hot water is generally delivered at temperatures up to 130°C (266°F) and returned to the plant at around 40–80°C (104–176°F). There are obvious economies of scale at most stages in the process: pellet fuel price, DHN installation, high levels of technological sophistication and operator skills. These would be difficult to achieve in smaller installations. Where there is no existing DHN, however, the initial costs of installation are high.

References

Niininen, H. Kankaanpää K. and Carbon Managers Network of Fortum (2006) *Fortum and Climate Change*, Fortum, Espoo, www.fortum.com/gallery/AboutFortum/Fortumand Climate Change Sep 2006.pdf (accessed 24 November 2009)

Sources of Further Information

Country codes and regulations

The codes and regulations surrounding energy efficiency and heat engineering vary considerably from country to country and region to region. They are also constantly changing. Below we list some of the most useful sites for updated information:

www.energycodes.gov/ A US Department of Energy website which includes an online code compliance tool – the RES*check* - developed to simplify and clarify residential code compliance with the Model Energy Code (MEC), the International Energy Conservation Code (IECC), and state-specific codes.

www.hud.gov Belonging to the US Department of Housing and Urban Development, this website has useful information on issues from housing to grants and has links to such matters in each state.

www.pprbd.com/PlanCkResEnergyCode.pdf At this web location, the Pike's Peak Regional Building Department have some useful information on building codes and heat loss requirements for family homes.

www.communities.gov.uk/planningandbuilding/buildingregulations/ The best site for access to UK information on building regulations, legislation, the planning process and even a code for sustainable homes.

www.planningportal.gov.uk/england/professionals/en/4000000000503.html Arguably the principal source for details on English building controls for wood pellet stoves and boilers, the UK government's planning portal deals with all wood fuel as if it were the same as any other solid fuel under Part J (Heat Producing Appliances). This site allows for downloading the full document which contains useful text and diagrams, as well as references and links relevant to air supply, ventilation, flue construction and design, the importance of separating combustible material (e.g. the wood pellet store) from the combustion and flue, debris collection, plus an appendix with methods for checking compliance with the requirements of solid fuel appliance regulation (J2).

www.direct.gov.uk/en/Employment/HealthAndSafetyAtWork/DG_4016686 and www.hse.gov.uk/fireandexplosion/workplace.htm both deal with the main relevant UK legislation on health and safety at work.

Wood pellet books, handbooks and websites

Books and handbooks

Solar Domestic Water Heating, Chris Laughton, Earthscan www.earthscan.co.uk (in the same series as this book)

Planning and Installing Bioenergy Systems, DGS and Ecofys, Earthscan www.earthscan.co.uk

Wood Fuels Basic Information Pack, Benet Bioenergy Network, Energi Dalen and Jyvaskyla Polytechnic, 2000.

Wood Pellet Heating: A Reference on Wood Pellet Fuels and Technology for Small Commercial and Institutional Systems, Massachusetts Divisions of Energy Resources, 2007.

Low Carbon Heating with Wood Pellet Fuel, a report for the Pilkington Energy Efficiency Trust; well illustrated and useful introductory pamphlet.

English Handbook for Wood Pellet Combustion, produced by PelletsAtlas, Force Technology and the National Energy Foundation (UK) under the EC's Intelligent Energy Europe programme, there are over 80 pages introducing wood pellet technology, 2009.

Wood Pellets, a booklet by Ralph W. Ritchie, this is written in an informal and anecdotal style to provide a reasonably good introduction to pellet stove use, Oregon (US), 2004.

Wood Pellet Boilers: A Guide for Installers, a booklet produced by Powys Energy Agency (Wales, UK) in conjunction with PROWE, a Finnish-led EC Altener project, 2003.

Biomass Heating: A Practical Guide for Potential Users (CTG012), The Carbon Trust, UK, 2009. www.carbontrust.co.uk/publications. This UK Carbon Trust introductory guide provides a short introduction to biomass heating and the key considerations for installing a successful biomass system.

Biomass Fact Sheet, Sustainable Energy Ireland.

Web links to wood pellet technology organizations and information

Wood pellet organizations in North America

www.pelletheat.org The Pellet Fuels Institute is a trade association located in Arlington, Virginia, USA. Their site provides information on wood pellets for residential and commercial applications as well as links to member pellet manufacturers in the US and in Canada.

www.woodpelletfuel.org/pellet_fuel_manufacturers/ This website offers a local directory for pellet and pellet boiler suppliers plus links to most major stove and pellet fuel manufacturers in the US.

www.pellet.org BC Pellet Fuel Manufacturers Association is a trade association located in Prince George, British Columbia, Canada. There is information on pellets and links to members.

www.greenenergyresources.com Green Energy Resources (formally New York International Log & Lumber Co.) is a green bio-energy supply company, with background information on the importance of biomass plus some industry links.

www.pelletsystemsconsult.com Pellet Systems Consulting is a North American consulting company that helps clients determine the feasibility of producing all types of densified products. PSC specializes in lowering the cost of producing wood fuel pellets, but their website has some useful general wood pellet information.

www.reap-canada.com Resource Efficient Agriculture Production (REAP) is a Canadian non-profit organization dealing with food, fuel and fibre. This website contains a database of reports, some of which mention wood pellets. This website is in English and French.

www.wood-me.com/timber_comps/1/107/9/258/2/Pellets_wooden_briquettes/ The 'Wood-me' website is an international business-to-business marketplace with Russian and Eastern European focus on the wood industry, including raw wood and wood-related products. This website features an online directory of wood pellet industry contacts.

www.woodfuel.com Mainly for large-scale US wood resource purchasers, this website has sections and links for pellet industry.

Wood pellet organizations in Europe

www.swedishenergyagency.se Swedish Energy Agency website with information on pellets and other biomass, plus member contact links.

www.pelletsindustrin.org Pelletsindustrins Riksförbund is the Swedish National Association of Pellet Producers; it has 12 member companies. English language version.

www.svebio.se Website for the Swedish Bioenergy Association (Svebio); it lists members by category of industry sector or service.

www.pelletsverband.at Pelletsverband Austria is an Austrian trade association with members among wood pellet manufacturers and suppliers and dealers of wood pellet heating equipment. German language only.

www.ecop.ucl.ac.be/aebiom/ The European Biomass Association members are national biomass associations across Europe.

www.bios-bioenergy.at/en/pellets.html One of Austria's top centres of excellence relating to wood pellet technology, their site provides information on pellets and downloadable publications.

www.pelletsatlas.info/cms/site.aspx?p=9127 This web page provides a range of excellent links for pellet and biomass information, standards and industry contacts.

www.esv.or.at ESV (the Energy Agency for Upper Austria) has a particularly proactive pellet promotion strategy, responsible for both the European Pellet conference and the World Sustainable Energy Days, possibly the best annual energy exhibition ever put together.

www.sh.slu.se/indebif/ INDEBIF is an EC ALTENER project aiming to expand knowledge about densified biomass fuels and to establish a European network for these. A database of wood pellet manufacturers in Europe is included.

www.propellets.at This site is provided by the Austrian Association for the Promotion of Pellets, which is dedicated to advancing the use of wood pellet heating; there are links to pellet producers and suppliers, as well as stove and boiler manufacturers.

www.britishbiogen.co.uk UK's British Biogen Trade Association site incorporates the Bioenergy Pellet Network and some useful information and links.

www.pdf-search-engine.com/wood-pellets-pdf.html This web page offers a useful research tool with links to several sites that have information on wood pellet technology.

www.soliftec.com The UK-based Solid Fuel Technology Institute site has a search engine with some useful links to information on pellets and combustion technology.

www.woodfuelwales.org.uk/wood-pellets.php The UK-based Woodfuel Wales website has introductory information on wood pellets and wood pellet combustion technology.

www.nef.org.uk/logpile/pellets/introduction.htm The National Energy Foundation (England) provides introductory information on wood pellets as well as pellet stoves and boilers through its Logpile website.

www.energysavingtrust.org.uk/housing The UK Energy Saving Trust website has information and publications on insulation, heating and ventilation for the home.

www.biomassenergycentre.org.uk This Biomass Energy Centre's website presents UK information relevant to biomass-derived solid, liquid and gaseous fuels and associated conversion technologies.

www.thebioenergysite.com This site features news and links on bioenergy across the world.

www.bioenergy.org.nz The Bioenergy Association of New Zealand (BANZ). Members include sawmillers, wood processors, energy suppliers, energy researchers, consultants, manufacturers and investors.

Useful software available online

A variety of quite useful and sometimes nationally approved software can be obtained, often free or on a free trial basis, from a variety of sources in Europe, the US and the UK.

www.retscreen.net/ang/version4.php RETSCREEN offers free online and downloadable software specifically designed for use with renewable project development, including biomass and wood pellet installations.

www.nher.co.uk This is the UK National Energy Foundation site with software for domestic energy efficiency ratings.

www.elmhurstenergysystems.co.uk/software-programmes.html Elmhurst Energy Services offer online software for assessing the energy performance of both new and existing dwellings.

www.eia.doe.gov/neic/experts/heatcalc.xls An excellent heating fuel comparison spread for North American markets, comparing pellet as a heating fuel with wood, fuel oil, natural gas and kerosene.

www.americanenergysystems.com/fuel-calculator.cfm US American Energy Systems offer a free online fuel calculator which compares corn, gas, wood pellets, electricity and oil.

www.pelletheat.org/3/residential/compareFuel.cfm An online home heating fuel costs comparison chart for North America.

www.pprbd.org/plancheck/Heatcalcpublic-v2.xls This is Pike's Peak's online heat loss calculator, very useful for North America.

www.ouellet.com/heat-capacity-calc-usa.aspx A Canadian online freely available software for heat loss and boiler sizing.

www.soliftec.com/fuelcost.xls This is the Solid Fuel Technology Institute's free online fuel and carbon calculator which compares fuel consumption and carbon emissions in the UK market for natural gas, heating oil, LPG, anthracite coal, wood logs, wood pellets and electricity.

www.builditsolar.com/References/energysimsrs.htm#Simulation A great link to a variety of US heat loss calculators, some useful for biomass and solar installations.

Courses and events

www.wsed.at The European Pellet Conference, held alongside the World Sustainable Energy Days conference and exhibition, is presented annually in Wels, Austria, and is the best-established wood pellet specialist event of its kind.

www.wood-pellets.com This site is dedicated to the *Spring Biofuel Congress*, St Petersburg, Russia (a great annual event), and the Russian *Bioenergy International* magazine.

www.cat.org.uk The Centre for Alternative Technology (Wales) provides information, services, education and training courses in a range of renewable technologies, including wood pellets and other biomass applications.

www.carmen-ev.de CARMEN, the 'Coordinating Office for Renewable Raw Material', is located in Bavaria, Germany and also a wood pellet training centre. This website is available in English.

www.renewableenergyworld.com *Renewable Energy World* is a magazine and online source of information on renewable energy technology, events and courses.

www.organicenergycompany.co.uk The Organic Energy Company is the sole distributor for ÖkoFen automatic wood pellet systems in the UK and Ireland; it has a network of well-trained installers and runs regular installation training courses.

www.greendragonenergy.co.uk Green Dragon Energy (UK and Berlin) offers training, books and DVDs on renewable technologies.

Manufacturers and suppliers of wood pellet burners and of equipment for the manufacture of wood pellets

The list of manufacturers and suppliers of wood pellet technology below is a selection of those companies who have in some way assisted in the production of this book. Any attempt at an extensive list would soon be out of date. To contact a wider range of manufacturers and suppliers we suggest identifying them via the relevant regional pellet technology associations and information-dispensing organizations.

www.ecobusinesslinks.com/wood-pellet-burning-wood-pellet-boilers.htm This is perhaps the best web page for links to manufacturers of wood pellet boilers in North America and Europe. It also has links to other sources of information on pellets and other biomass.

www.organicenergycompany.co.uk The Organic Energy Company is the sole distributor for ÖkoFen automatic wood pellet systems in the UK and Ireland; it has a network of well-trained installers and runs regular installation training courses.

www.greendragonenergy.co.uk Green Dragon Energy (UK and Berlin) offers training, books and DVDs on renewable technologies.

www.veryefficientheating.co.uk The Very Efficient Heating Company (UK) provides solar and biomass heating installation services.

www.okofen.at, www.oekofen.co.uk, www.oekofen-usa.com ÖkoFen is a specialist wood pellet boiler manufacturer based in Austria but with UK and US agents.

www.froeling.com Fröhling is a well-established Austrian biomass boiler manufacturer that makes a range of wood pellet stoves and boilers.

www.viessmann.com Veissmann is a heating engineering company which manufactures a range of pellet boilers and has worldwide representation.

www.koeb-holzfeuerungen.com/kus_tree//powerslave,id,1,nodeid,1,lang,EN.html Köb offers modern heating technology across Europe for firewood, shavings, pellets and wood chips in the 35–1250kW range.

www.highlandwoodenergy.co.uk Highland Wood Energy is Scotland's leading wood heating specialist, dedicated to the provision of practical solutions for wood-fuelled heating, including wood pellet stoves and boiler installations.

www.americanenergysystems.com A well-established US-based company which supplies wood pellet stoves.

www.pelletstoveusa.com Pellet Stove USA delivers and installs quality pellet stoves in Maine Massachusetts and New Hampshire.

www.armstrongpellets.com A Canadian-based company selling pellets and also pellet stoves.

www.pelletstove.com A Dell-Point Technologies division – Alternate Energies for Today – website with information on pellets and pellet stove sales.

www.ashwellengineering.com Ashwell Engineering Services Limited is a UK-based manufacturer and installer of Green-tec range wood pellet boilers with outputs from 15kW to 1.2MW.

www.eco-link.co.uk Eco-Link install boilers and other renewable technologies in the UK.

www.janfire.com Janfire is a Swedish manufacturer and installer of high-specification pellet boilers from 20kW to 3 × 600kW.

www.jones-nash.com Jones-Nash is the sole importer for ETA Heiztechnik (Austria) and Osby Parca (Sweden). Stoves and boilers range from 15kW (domestic) to 5MW (large-scale commercial).

www.brites.eu Balcas wood pellets are marketed as *brites*. They produce 155,000 tons of wood pellets in Ireland and Scotland every year from sawmill by-product.

www.blazersfuels.co.uk North Wales-based wood pellet manufacturer based at sawmill (pellets are made of 100 per cent virgin timber wood chippings and contain no additives).

Index